Carl Hering

Recent Progress in Electric Railways

Being a Summary of Current Periodical Literature Relating to Electric Railway

Carl Hering

Recent Progress in Electric Railways
Being a Summary of Current Periodical Literature Relating to Electric Railway

ISBN/EAN: 9783337203993

Printed in Europe, USA, Canada, Australia, Japan

Cover: Foto ©berggeist007 / pixelio.de

More available books at **www.hansebooks.com**

RECENT PROGRESS

IN

ELECTRIC RAILWAYS.

———

" The more a science advances, the more it becomes concentrated in little books."—LEIBNITZ.

WORKS BY THE SAME AUTHOR.

PRINCIPLES OF DYNAMO-ELECTRIC MA-
CHINES AND PRACTICAL DIRECTIONS
FOR DESIGNING AND CONSTRUCTING
DYNAMOS, *with an Appendix containing several
Articles on Allied Subjects, and a Table of Equivalents
of Units of Measurements. Sixth Thousand, cloth,
279 pages, 59 illustrations,* - - . - $2.50

UNIVERSAL WIRING COMPUTER. *For De-
termining the Sizes of Wires for Incandescent
Electric Lamp Leads and for Distribution in General,
Without Calculation, Formulæ or Knowledge of
Mathematics, with some Notes on Wiring and a
Set of Auxiliary Tables. Cloth,* 44 *pages,* - $1.00

PRACTICAL DIRECTIONS FOR WINDING
MAGNETS. *Cloth,* 63 *pages,* - - $1.25

TABLES *of Equivalents of Units of Measurements.
Flexible covers,* 11 *pages,* - - - - .50

---

*Copies of these books will be mailed to any address in the
world, postage prepaid, on receipt of the price.*

The W. J. JOHNSTON COMPANY, Ld.,
TIMES BUILDING, NEW YORK.

RECENT PROGRESS

IN

ELECTRIC RAILWAYS

BEING

A SUMMARY OF CURRENT PERIODICAL LITERATURE
RELATING TO ELECTRIC RAILWAY CONSTRUC-
TION, OPERATION, SYSTEMS, MACHINERY,
APPLIANCES, ETC., COMPILED

BY

CARL HERING,

Author of "Principles of Dynamo-Electric Machines,"
"Universal Wiring Computer," etc.

———

NEW YORK:
THE W. J. JOHNSTON COMPANY, LTD.,
167-176 Times Building.
———
LONDON :
WHITTAKER & COMPANY,
Paternoster Square.
———
1892.

PREFACE.

In the earlier days of the development of electrical engineering the current literature was so small in amount and the branches so few that the electrical engineer who wished to do so found little difficulty in keeping up with the progress in every department, and in making sufficient notes or abstracts to permit of reference to the various sources in which information of interest to him had been published.

The progress in this active field during the past few years has been such, however, that the volume of current electrical literature, as well as the number of branches of electrical engineering, is becoming so great as to render it almost if not quite impossible for one to keep satisfactorily informed of all that is going on in the different departments, much less to take the time to keep notes and records that will enable him afterward to refer back to an article or description that he knows he has read, but does not remember where.

The numerous electrical and other technical journals, which give such a large amount of valuable information, must cover so many fields that their bound volumes are not convenient records in any one branch, besides being of necessity large and bulky, and therefore not handy for connected reading or for subsequent reference.

As a consequence of this it is thought that a series of little volumes in clear, legible type, well printed on good paper and substantially bound in cloth, containing a *résumé* to date of what is of value in current electrical literature, will appeal to those who do not wish to lag in the march of progress. Should this first volume meet with the success which it is hoped it will merit it is the intention to extend the series by issuing other volumes embracing all the important branches of electrical science, and also adding to these from time to time as the advance in the respective branches shall warrant.

Each volume is intended to contain a classified summary up to date of the literature on that particular branch or department from the time of the publication of the preceding volume on the same subject, to serve not only as a record for reference, but also as a book containing the latest information obtainable on the subject. Articles which have been published will be given in full or in abstract, or only by reference to the source, depending on their importance, length, etc.

"Current Progress in Electric Railways," being the first of the series, has not had as much time devoted to its preparation as the compiler would have liked, and moreover most of the material for this volume has been taken from the columns of one journal. In subsequent volumes it is intended to cull from all the prominent electrical and other engineering journals, foreign as well as American, and to so arrange and connect the matter that the work shall have rather the character of a logical summary of recent progress than that of a mere compilation. It is aimed to make these books standard works of their kind. References by foot notes and otherwise will be made to articles of a character too special for the scope of these volumes, yet of value in special researches or to special students.

It is understood, of course, that no responsibility is assumed for statements compiled, or that their insertion necessarily implies any endorsement. A work of this kind must include some descriptive matter relating to new inventions, devices, etc., but these will be regarded from a technical rather than from a commercial standpoint.

Any assistance or suggestions tending to enhance the usefulness of these volumes are solicited and will be cordially appreciated.

CONTENTS.

RECENT PROGRESS IN ELECTRIC RAILWAYS.

CHAPTER I.

HISTORICAL NOTES.

During the past year some interesting information has been added to our present knowledge of the
history of the electric railway, including also some
compilations, abstracts of which are given below
in order of date.

1835-37.—Mr. Frank L. Pope—who may with
right be called the discoverer of the inventor of
the electric railway—found quite accidentally that
a certain Thomas Davenport constructed a model of
a small electric railway as early as 1835, which
model still exists. In a very interesting lecture on
electric railways (reprinted in full in *The Electrical
World*, January 31, 1891, page 81) Mr. Pope remarks
as follows:

"In the expectation that electricity would soon
be the universal motive power, the problem of
its utilization was, during the next ten years,
(1831–41) attacked with great zeal by a considerable
number of inventors. The earliest, as well as the
most meritorious of these was a Green Mountain
village blacksmith, by the name of Thomas Davenport. Some time since, while preparing a lecture to
be delivered before the Board of Trade, in Spring-

field, Mass., I accidentally learned that this inventor had constructed a model of an electric railway in Springfield more than fifty years ago. This was undoubtedly the very first appearance of the electric railway in the history of the world. The work of this electrician, and even his very name, is almost unknown to the present generation of laborers in this fertile field.

"Time does not permit the rehearsal of the detailed story of his struggles, his failures, and his successes. During the six years between 1835 and 1841 he built more than 100 operative electric motors, scarcely any two of which were of the same design, and which varied in size from a small model up to an engine capable of driving a rotary printing-press for twelve hours in succession. In December, 1835, he exhibited in Boston the model of an electric locomotive and circular railway which he had built in Springfield, and a similar, though more finely finished, model which was built by him the following year is still in existence in a good state of preservation. He published in 1840, now more than fifty years ago, in the city of New York, a weekly journal called the *Electro-Magnet and Mechanics' Intelligencer*, which was printed upon a press driven by an electro-magnetic engine. Some of his motors, constructed as early as 1837 and 1838, were admirably designed, and in essentials, even to the shunt-winding, differed but little from some of the most advanced and successful types in use at the present day."

In a paper read by the same author before the American Institute of Electrical Engineers (reprinted in *The Electrical World*, March 21, 1891, page 228) he describes the inventions of Thomas Davenport, from which we make the following extract:

"The year 1837 marked a very important era in the history of the industrial development of electricity. During that year two of the most extraordinary inventions of the present century made their advent in this city (New York), the electric telegraph of Prof. Samuel F. Morse and the electric motor of Thomas Davenport. The motor came first. Davenport, a self-taught Vermont blacksmith, who had invented and constructed his machine in a remote country village in a crude form, as early as 1834, came to New York in February, 1837, bringing with him some of the machinery which had been made for the purpose of exhibition, by himself and Ransom Cook, with a view of enlisting capital to build a large motor." Mr. Pope found that several of the original models were still in existence, one of these being shown in the following cut (Fig. 1), and was exhibited in operation by him at his lecture, even with the original three-cell Grove battery of pint cups. It is a circular railway, $2\frac{1}{2}$ feet in diameter, with the locomotive traveling on it. He thinks that there is no doubt that that model was built in the early part of 1837, possibly as early as 1836. The locomotive has a fixed field magnet below, and a revolving armature above, through which the current is reversed twice in every revolution. "So far as I have been able to discover, that is a combination found in every practical motor to-day, which Davenport was the first to make known and to use in 1834. The motor is connected to the axle by bevel gear. The field magnet and armature in the model are connected in series. In another model of 1837, they are connected in shunt."

Further information about the electric motors of Davenport will be found under the history of motors in another volume of this series.

Mr. Riley Bowers writes, concerning Davenport's

FIG. 1.—DAVENPORT'S ELECTRIC RAILWAY.

model of his electric railroad: "My father told his family in my presence that when Mr. Davenport arrived in New York he was offered $250,000. He would not take it, but took his model to England and set it running. Michael Faraday was well pleased with it, but after it had been running some time it occurred to him to try its power. He took a broom that was in the room, put it against the fly-wheel, and stopped the motor. After that Mr. Fara-day refused to invest, or recommend it to others. So

Mr. Davenport had to bear all the expense, and received nothing for his trouble."

1850.—Thomas Hall is said to have made an electric locomotive that was "the marvel of the time."

1851-52.—In this year an electric locomotive was constructed by Prof. George G. Page, using batteries of the Grove type, for which experiment Congress voted $50,000; it was tried on the Baltimore and Ohio Railroad, and was said to be a partial success. Regarding this experiment, Mr. Pope says: "It was a noteworthy achievement, but developed no new practice, nor did it contribute anything to the ultimate solution of the electric railway problem. Many other promising experiments were made, but all resulted in failure, until the very name of the electro-magnetic motor became almost a synonym for 'humbug.' "

1855.—Mr. A. M. Tanner published the following facts in the *London Electrical Review* (reprinted in *The Electrical World*, December 19, 1891, page 453): He states that Major Alexander Bessolo, now in Turin, Italy, and at that time lieutenant in the artillery of the Sardinian States, proposed an electric railway system in which an overhead conductor and the rails served as a circuit, and the rotary electric motor on a car was included in a traveling connection between said overhead conductor and the rails. In reference to the propulsion of cars on railways, the inventor states that the current is conveyed to the locomotive motor machine by means of the rails and by a conductor insulated from the ground and suspended in a manner analogous to telegraph wires. The inventor furthermore states:

"Such a system of locomotion has many advantages, because the current can conveniently reach independent vehicles or those united in small trains. Also, encounters or collisions on the same track are rendered impossible, because the same current will never allow two trains to be present in the same section with a common conductor. Besides, the generators are so located that they can be controlled from the stations."

Mr. Tanner also states that "in the same year Chevalier Bonelli, inspector of telegraphs of the Sardinian States, invented his so-called locomotive telegraph system and worked it practically between Paris and St. Cloud, for which a French patent was granted on January 9, 1855. It showed what is known as the 'third rail conductor' for conveying an electric current to a translating device upon a car through the medium of a contact device sliding upon said third rail conductor. In this system of Bonelli, as well as in others devised about the same time by De Castro, Guyard, Du Moncel and others, for establishing a telegraph communication between a train and the station, it was proposed to use a conductor insulated from the ground, and the rails as a circuit, and obviously in all these systems the translating device was in derivation between the insulated conductor and the rails. Consequently, when Bessolo stated that he proposed to use an overhead wire and the rails of a railway track as the circuit of a stationary generator of electricity, he obviously had the intention of completing the circuit through a traveling connection between the overhead conductor and the rails. This naturally placed the electric motor on the car in a branch or 'leak' between the overhead conductor and the rails, just as is the case in all electric railways of the overhead conductor and trolley type."

1860.—Hall is said to have exhibited an improvement on his locomotive of 1850, and to have run it upon a circular track.

1874.—Mr. C. J. Van Depoele was experimenting with electric motors in Detroit, and it occurred to him that trains of cars and even commercial street cars could be run by electricity; this was demonstrated to the satisfaction of his associates in various ways, but no public exhibition was made until 1883.

1877.—Mr. F. L. Pope writes: "The credit of priority in the invention of the electric railway in its modern form—that is to say, the moving electro-motor on the car connected by electric conductors with a stationary dynamo—I believe to be justly due to Stephen D. Field, now of Stockbridge, Mass., who was living in San Francisco in 1877. This was about the time of the introduction of the cable system of street railways, which was resorted to in San Francisco, for the reason that some of the grades were so excessive that the use of horse power was absolutely out of the question. Observing the operation of the cable road, it at once occurred to Field that electric power might be applied to the same purpose with as great or even greater advantage. At that time no dynamo-electric machines suitable for this purpose were be to had in the United States, and he accordingly ordered one from Europe for experimental purposes. It took a long time to make it, but at last it was completed and shipped to San Francisco by a sailing vessel. The vessel was wrecked on the voyage, and the machine went to the bottom of the sea. Not yet discouraged, he ordered another one, which eventually reached him in good order, and enabled him to commence his long-delayed experiments. He tried first an electric elevator, in which he was successful. In 1879, having exhausted his resources, he came to New York, bringing with him his plans, with which he hoped to enlist capital to continue his work. He laid these plans before me, and being impressed with the entire practicability of his scheme, I gave him every encouragement in my power. The plan which he had devised contemplated the inclosing of the conducting wire in a conduit beneath the street. He was not successful in obtaining sufficient means to properly develop his invention; he became involved in tedious, harassing, and expensive litigation with wealthy corporations, and his health failed him at a critical time, so that for years he was incapacitated from active work; but the single railway which is in operation in this country to-day, and embodies his matured conceptions, is regarded by competent judges as in many respects superior to any yet brought before the public.

1882.—The same writer states: "In the summer of 1882, Dr. Joseph R. Finney exhibited in Allegheny, Pa., an electric street car, for which the current was supplied by a copper wire, about the thickness of a lead pencil, 15 or 20 feet above the street. A small trolley, fitted with grooved wheels, running on this wire as on a track, and connected with the car by a flexible conducting cord, served to convey the electric current from the suspended conductor to the motor. Some of the earliest of the successful lines in this country were arranged upon this plan. At a later date this was super-seded by the type of contact wheel now in general use, which runs underneath the wire and is mounted upon the end of a long yielding rod bearing a certain resemblance to a fishpole, supported upon the roof of the car."

1883.—Leo Daft operated a full-sized passenger car over the Mt. MacGregor Railway at Saratoga. Van Depoele exhibited a car in operation in Chicago. Field's electric locomotive was exhibited at the Exhibition of Railway Appliances in Chicago, transporting in the aggregate 27,000 passengers.

Mr. Eugene Griffin states: "When the Chicago elevated railway was under consideration, it was proposed to demonstrate the feasibility of utilizing electricity as a motive power. A track 400 feet in length was built, with a 5 per cent. grade in the centre. One car was equipped with a 3 horsepower motor, and ran for several weeks with considerable success, carrying crowds of people. This was in February, 1883. In the same year an elevated railway car was operated electrically at the Chicago Inter-State Fair. The car was suspended from the truck instead of being mounted on it; and was in operation during the entire exposition—some fifty days.

"On July 27, 1884, an electric car was running scheduled trips over a mile track of the East Cleve-

land Street Railway Company, in Cleveland, O.
This was the first electric car in regular operation
on a street railway track in the United States. The
motor was placed between the wheels and supported
from the car body, and geared to the axles by belts
of spring wire cables. The current was conveyed
to the car by the conductors supported on insulators
in a small wooden conduit, and connection made
with the conductors by means of a plow extending
through the slot to the conduit. This was the initial
installation of the Bentley & Knight system. The
road was given up in 1885.

"During the Toronto Annual Exhibition in 1884,
an electric railway some 3,000 feet long was operated from the entrance to the grounds to the main
building. This was a conduit road, and the wires
carried a potential of over 1,000 volts without accident. A 30 horse-power electric locomotive was used
hauling trains of cars."

1885.—Mr. Pope writes: "It is only six years
ago that an electric street railway was put in
actual commercial service in the United States
for the first time. This was in the city of Cleveland, O. To-day considerably more than one-third
of the total street railway mileage of the United
States is either operated by electric power, or
contracts have been entered into for the substitution of an electrical equipment for that now in use
at the earliest possible moment."

1886.—Mr. F. J. Sprague writes: "The development of electric traction is unequaled in the
industrial history of the world. In 1886 a list of
twelve or thirteen comprised all the electric roads

In operation, and this included every electrical road in the world, whether operated by the split tube, the side rail, the traveling trolley carriage on an overhead wire, the centre rail, or by a conduit or storage battery. There was little similarity in these different plans, but they all served to show that electricity, in a more or less effective way, could propel a car."

1887–88.—Mr. Griffin states: "After various experiments, the road in Allegheny City was begun in the summer of 1887. The cars were started during the winter of 1887–88, although the road was not formally opened to traffic until February, 1888. Four cars were furnished to this road, which, I believe, are still running. On the lower end of the road was a mile of double track conduit, which was continued by an overhead system of about five miles. The conduit was on a long grade of about 12 per cent. Over-running trolleys were used with the overhead system. The conduit was in operation for two years or more, but has now been taken up and replaced by the overhead system.

"It was not until 1888 that the electric railway became a practical commercial success. I fix the date at 1888, as it was in that year that Bentley & Knight opened the Allegheny City road to regular traffic; that the Sprague company equipped the Richmond road, and the Thomson-Houston company installed the Eckington & Soldiers' Home Road in Washington. It was in 1888 that railway officials began to realize the possibilities of this new active force, that the great West End system of Boston adopted electricity to the exclusion of cable, and that orders began to flow in upon the electric companies for street car motors to such an extent as to soon make the manufacture of such motors one of the leading branches of the electric industry.

"Previous to 1888 electric motors had been used on several roads. Some of these roads were doing well and have been prosperous since; but to the public these were experiments on a comparatively small scale, and did little to inspire general confidence. The early inventors found it difficult to secure adequate financial backing; orders were few and business unprofitable. The stronger companies, which took up the work in 1888, had the organization and capital necessary to achieve success.

"In the fall of 1887 Frank J. Sprague contracted for the electric equipment of the Union Passenger Railway, at Richmond, Va. This was an important road in a large city, and Mr. Sprague's undertaking was the most ambitious effort in this direction up to that date. It is worthy of note that Sprague's original intention was to use motors with but one reduction, but he was forced to abandon this idea, as none of the electrical companies of that date were able to produce single reduction motors. The motors used at first were too light for the work, the copper brushes scored the commutators badly, and were rapidly consumed. Nevertheless, Mr. Sprague persevered despite all obstacles, and in 1888 the road was running with so much success that it was one of the object lessons which induced Henry M. Whitney and his brother directors of the West End Street Railway of Boston to adopt electricity as a

motive power when they were already far advanced
in the plans for cabling their system."

Regarding this road in Richmond, Mr. Pope writes:
"Lieut. Frank J. Sprague designed, carried out,
and completed the first installation of electric railroading on a large scale in the world, in Richmond,
Va., in 1888. Laboring under enormous difficulties
and drawbacks, Lieutenant Sprague succeeded, by
the completion and operation of this plant, in establishing beyond peradventure the future supremacy
of the electric street railway, and many of the char-
acteristic features at that time designed and introduced by him have practically become standards in
the modern system, and are found in nearly every
one of the thousands of cars now in service."

Mr. Sprague himself writes regarding this road:
"On the 8th of February, 1888, there was opened
for traffic, under the Sprague system, a road at
Richmond, Va., which presented conditions of
length, grade, curves, road-bed, and number of cars
to be operated, which, if successfully overcome,
would mark a new era in the development of electrical railway traction. The conditions, while not
perhaps now seeming remarkable, were then considered insurmountable, not only because of difficulties relating to street car service itself, but also the
electrical and mechanical ones. The length was
from 11 to 12 miles. There was a straight grade
of 10 per cent.; there were grades in curves of
7 and 8 per cent.; there were twenty-nine curves,
and some were as low as 27 and 30 foot radius.
The roadbed was of an execrable character.
Thirty cars had to be operated at one time from

a common station, and some of them four miles
away from the station. That road had its vicissi-
tudes, but its victories as well. Forty cars were
operated, and no less than twenty-two simultaneously at one end of the line. The electrical and
mechanical features, hastily designed and crudely
constructed, were a radical departure from the previous work."

A paper by Mr. Griffin contains the following
summary: "As nearly as can now be ascertained,
the following electric roads were actually in operation on January 1, 1888:

Roads.	System.	Location.	Miles.	No. of mtr. cars.
Union Passenger Ry. Co........	Daft	Baltimore, Md......	2.00	3
Windsor Electric Ry............	Van Depoele opp.	Detroit, Mich.......	1.25	2
Appleton "	"	Appleton, Wis......	5.50	5
Port Huron "	"	Port Huron, Mich...	2.75	4
Highland Park................	Fisher	Detroit, Mich.......	3.25	4
Scranton Suburban road.......	Van Depoele.....	Scranton, Pa......	5.00	12
Los Angeles Electric Ry. Co.....	Daft.............	Los Angeles, Cal...	5.00	6
Lima Street Ry. and Motor Power Co.................	Van Depoele.....	Lima, Ohio.........	4.00	8
Columbus Consolidated Street Ry.	Short	Columbus, Ohio.....	1.00	2
St. Catherines Street Ry. Co....	Van Depoele.....	St. Catherines, Ont.	7.00	12
Seashore Electric Ry. Co........	Daft.............	Asbury Park, N. J..	4.00	18
San Diego Street Ry. Co.........	Henry............	San Diego, Cal......	3.00	9
E. Harrisburg Passenger Ry. Co.	Sprague..........	Harrisburg, Pa......	4.50	10

"A total of 13 roads, 48.25 miles of track and 95
cars.

"On July 1, 1891, there were 354 roads in actual
operation, with 2,893 miles of track equipped electrically, and 4,513 motor cars. Such has been the
growth of three and a half years. On January 1,

1889, the first electric car was started in Boston with the Sprague system, and later the Thomson-Houston system was adopted."

CHAPTER II.

DEVELOPMENT AND STATISTICS.

The development of electrical railway engineering has been so rapid, that it is perhaps without a parallel in any other industry. It was only a few years ago that an electric railway was still considered an experiment, while at the present time horse-car lines and cable lines are not only being rapidly replaced by electrical roads, but it is not likely that many new horse-car lines will be started, and there are many cases where electric roads have been built where horse-car lines would hardly have been considered feasible.

The rapid development is illustrated very well by the following extract from a paper read at a street railway convention: At a meeting of the American Street Railway Association, held in Washington in 1888, almost the sole subject of discussion was the question of improved motive power for street railway service. Every manager and superintendent present realized that the demands of traffic were rapidly outgrowing the existing methods, and many of them had begun to look forward to the application of electric power as the only possible solution of the problem. As one delegate after another reported his experiences and observations in respect to electric traction, a member who had listened to it all with evident impatience, arose and entered his solemn protest against all the new-fangled talk about electricity. He said that he had come there upon the assumption that it was a horse railroad convention, in order to get information about horses. He wanted to know how to shoe them, how to feed them, and how to work them to the best advantage. And as he continued to complain, it was evident that in more than one place in the assemblage his ideas met with a sympathetic response. In September, 1890, the same association held its annual meeting at Buffalo, and, in his opening address, the president, Thomas Lowry, of Minneapolis, made the following significant remark: "I am so thoroughly convinced that electricity is the coming power for street railways (except on very heavy grades, where the cable is best suited), and that it will prove so effective as a means of rapid transit for cities, that I believe that this is the last convention that will ever seriously consider horses for the operation of street railways."

The development of the street railway has as much, or perhaps more, to do with the growth and prosperity of towns and cities than any other one thing, as the transportation of people through the city is most intimately connected with the social and business life of the people. The daily loss of time to the people of any city where horse-car lines are run at four to six miles an hour, when compared with electric roads running at six to twelve hours, is far greater than one would at first suppose. The effect of rapid transit on the growth of cities is shown very well in the following extract from a paper by Mr. Griffin, from which will be seen what an important effect rapid transit has on the increased value of property in and about towns and cities: "Let us assume that a man can allow thirty minutes morning and evening for his car ride, paying five cents for each ride. At the rate of six miles per hour, fast for horses, he has a radius of three miles and an area of 28¼ square miles within which to select a home. At the rate of nine miles per hour, slow for electricity, he has a radius of four and a half miles, and an area of 63½ square miles within which to select a home. An increase of only three miles per hour in rapidity of transit doubles the available residence area without increasing the time or expense of the laborer in going to and from his work."

In dealing with the problem of street railways for any city, one should not consider only the present population, but prepare also for the great increase which is certain to come in most of our American cities. There are 74 cities in the United States which have a population of over 40,000. In these the average increase of population during the last 10 years was nearly 47 per cent. While this rate of increase may not continue at that high figure, the growth is however reasonably sure to be very great for quite a number of years.

Among the details in electric railroad engineering, there are a few features whose development were especially marked during the past year. Chief among these was the reduction in the gearing. Formerly, a double reduction gearing and a high armature speed in the vicinity of 1,000 revolutions were almost universally used; at present most of the motors introduced have but a single reduction, the armature speed having been reduced to about one-third, while the gear wheels are now usually run in cases filled with oil or grease, which has greatly reduced the noise and the wear and tear of the gearing. An attempt has also been made to do without any gearing by connecting the armature directly to the car axle, forming the so-called gearless motor. Although reports appear favorable, it cannot be said that it is likely to replace the single reduction motor. It is claimed that the necessary loss in efficiency of these gearless and single reduction motors over a double reduction is not as great as the gain in having less gearing.

From reports received from different companies, it appears that a perceptible improvement has been made within the last year in the mortality of armatures, and it is likely that a still further improvement will follow the introduction of the single reduction and gearless motor. It was stated that at least three armatures were damaged to one field coil rheostat or switch box. The general adoption of the Gramme armature has also decreased armature repairs. In the stations the tendency has been to use larger dynamos, large and more economical compound and triple-expansion engines, and a better proportioning of the plant to the average power required. Another improvement has been made in the substitution of long cars for short ones, which is accompanied by the use of two pivoted trucks to a car in place of the usual rigid truck and frame used with the ordinary 16 foot cars.

In comparing the relative merits of cable railways with those using electrical power, it seems to be the almost universal opinion that the former are better for very heavy grades and where extremely heavy traffic is to be handled, and probably also where distances are very short.

The comparative merits of the two systems are shown in the following extracts from an article by Mr. J. C. Henry, in which he states that arrangements had been completed in Boston for running cable railways. After investigation, the company changed their plans and have now introduced electric power. The same was true of the Minneapolis and St. Paul railways, while in Omaha the cable plant had already been introduced, but was abandoned in favor of electricity. He adds: "To build a cable road on Broadway (New York) of the most approved pattern requires excavation 4 feet deep and about 15 feet wide along the entire street; many gas and water pipes would have to be removed; the street, in all probability, would be torn up and partially blocked for a year, as the work can only be carried on during the building season. The residents know full well what this means—with a soil so polluted with gas that a single paving block cannot be removed without making the atmosphere offensive. Cable conduits mean two open sewers the entire length of the street. They are certainly objectionable from a sanitary point of view. Electric cars can be stopped much quicker than cable ones, for the reason that the propelling force can be instantly reversed. Cable cars necessarily run at a fixed rate of speed. Electric cars can be operated with the greatest certainty at any desired speed. An accident to an electric car means it must stop; accidents to grips or cables sometimes means the cars must go whether the road is clear or not. Imagine a heavy cable car going down lower Broadway with a snarl caught in its grip and no way of stopping it until word was sent to the power house to shut down the engines. Such accidents are of frequent occurrence on cable roads."

STATISTICS.

In connection with the development of this industry and its relative importance in the whole street railway business, the following statistics, compiled from various sources, may be of interest.

It is only six years ago that an electric street railway was put in actual commercial service in the United States for the first time. This was in the city of Cleveland, O. To-day considerably more than one-third of the total street railway mileage of the United States is either operated by electric power, or contracts have been entered into for the substitution of an electric equipment for that now in use at the earliest possible moment.

Mr. F. L. Pope stated: "The official returns show that during the year ending September 30, 1889, the 110 street railways in the State of New York carried over 686,000,000 passengers, or 100 times the total population. In New York city alone the surface and the elevated roads carried together about 400,000,000. In Boston 100,000,000 and in Philadelphia 150,000,000 passengers were carried. In fact, statistics indicate that the street railways of the United States carry something like twice as many passengers as all the steam roads, and moreover, it has also been found that the number of passengers increases from year to year in a much greater ratio than the population, which means, not only that more pecple ride, but that the same people ride more frequently each succeeding year."

Regarding the traffic in the city of New York, Mr. Sprague states: "The total annual passenger traffic—that is, the total number of persons carried in this city—has increased at the rate of over 140 per cent. in each period of 10 years since 1866, and is now something over 325,000,000. At the same rate of increase it would amount in 1890 to over 500,-000,000 and in 1900 to 1,225,000,000.

Mr. John N. Beckley states: "As many as 30,000 street cars, horse, cable and electric, are to-day (Sept., 1891) running upon the 8,000 miles of street railroads in this country. In these cars, and on these tracks, are carried as many as 3,000,000,000 of people yearly, or 50 times the entire population of the United States. When we consider that the number of people carried by all of the steam railroad companies in all of the States of this Union last year is estimated at less than 500,000,000, and that more people are carried on the street surface railroads in the city of New York, in a year, than are carried by all the steam railroads of the State in the same period, we come to have some conception of the immense importance to the people of the rapid, efficient and safe service of street cars in the rapidly growing cities and towns of this wonderfully prosperous country."

The new census bulletin shows that in four of the largest American cities the mileage of street railways of all kinds was nearly doubled between the years 1880 and 1889, the figures being 1,983 miles in 1880, and 3,150 in 1889. In December, 1889, 476 cities and towns in the United States had street railways. There is now scarcely a town of 5,000 inhabitants without one or more street railways.

The following very good abstract of the Eleventh United States Census appeared in the *Street Railway Journal* for January, 1892:

"The development of street railways during the decade lying between the tenth and eleventh censuses—a development both as to facilities and amount of business done—may certainly be counted as one of the most remarkable features of the whole comprehensive business of transportation. Looking first to the question of length, it is found that in 1880 there were 2,050 miles of street railways in operation, while in 1890 this number had risen to 5,783 miles, an increase in the ten years of 3,733 miles. This increase, remarkable as it is for the whole ten years, is still more remarkable when the decade is divided into two periods of five years each, for then it is seen that the most astonishing development has been during the last half of the ten years, and at a rate before unparalleled. The figures show that during the first five years the increase of mileage was 888 miles, while during the last half it was 2,845 miles. Looking for the cause of this extraordinary increase, it can readily be found in the introduction of electric roads. Of these roads, which on June 30, 1890, constituted nearly one-fifth of the number of street railways, none were in operation previous to the year 1886. In that year two electric railways commenced operations; in 1887 the number had increased to six; in 1888 to thirty; and in 1889 to fifty-seven, while during the first six months of 1890 no fewer than forty-nine new electric roads were reported. The development of cable roads has also largely assisted in this increased mileage, but not nearly to such an extent as the electric railway, while, as has been shown, the year 1886 was the year of inception of the electric road, the first cable road began to run in 1887. The increase by years is shown in the following table:

Years.	Total Length (Miles.)	Increase.	
		Miles.	Per cent.
1880....	2,050.16		
1881	2,150.09	99.93	4.87
1882	2,342.20	192.11	8.93
1883	2,506.14	163.94	7.00
1884	2,680.31	174.17	6.95
1885	2,938.29	257.98	9.93
1886	3,268.58	330.29	11.24
1887	3,890.22	621.64	19.02
1888	4,499.49	609.27	15.66
1889	5,285.11	785.62	17.46
1890 (6 months)	5,783.47	498.36	9.43
Ten years, 1880–1890		3,733.31	182.10

"Looking to the urban locality of increase, it is found that the most remarkable is in the smaller

cities, a fact that is plainly illustrated in the sub-
joined table:

ITEMS.	Length of Street Railways		Factor of Increase.
	1890.	1880.	
All cities...............................	5,783.47	2,050.16	2.82
Cities of more than 50,000 inhabitants..	3,205.59	1,584.16	2.02
Cities of less than 50,000 inhabitants...	2,577.88	466.00	5.53

"When the final computations were made it was
found that on July 1, 1890, the street railway companies of the United States in independent operation
numbered 789, and that these, repeating the previous figures, carried on their operations over 5,783
miles of street line, or over a total track length of
8,123 miles. On this length of line 32,505 passenger
cars were in use; the roads and equipment cost, all
told, $389,357,289, they gave employment to 70,764
men, and carried the astonishing total of 2,023,-
010,202 passengers. A good idea of the extent of
this traffic may be realized when it is stated that the
street railways of the United States carried last
year a number of passengers considerably greater
than the population of the globe, and when it is also
stated that the steam railways of the United States
with all their 157,759 miles of line and 25,665 passengers cars only carried during the same year 472,-
171,343 passengers, or 1,550,838,859 passengers less
than were carried by the street railways."

In September, 1891, the same journal published the

following summary, which is probably the most reliable one that has been published: "The total number of miles of street railways in the United States and Canada is 11,029. Of these 5,442 are operated by animal power, 3,000 by electricity, 1,918 by steam and 660 by cable. The total number of cars in the United States and Canada is 36,517, of which 25,424 are run by animal power, 6,732 by electricity, 3,317 by cable and 1,044 by steam. The total number of lines in the United States and Canada is 1,003, of which 412 use electricity and 54 cable. The number of horses is 88,114, mules 12,002 and steam motors 200. The diminution of the number of horses in one year was 28,681.

Mr. Sprague, in February, 1891, published the following summary: "The number of electric railways in the United States is 310; number of motors, 7,000; total horse-power, 175,000; the number of car miles run per day, 400,000; passengers carried annually, 1,000,000,000; the investment in horse-car lines, $58,000,000,; the investment in electric roads, $50,000,000; and that in cable roads, $49,000,000."

Mr. Watson, in October, gave the investment in electric roads as $75,000,000. Mr. Griffin gave the number of miles of electric railroads as 2,893, the number of cars on these roads 4,513, and the number of electric railroads 345. By way of comparison it may be interesting to note here that on January 1, 1888, the number of electric railways was 13, the number of cars 95 and the number of miles 48.25.

The *Street Railway Journal* published in October, 1891, the number of miles of track of street railways in the following cities:

Philadelphia	510	Kansas City	141
Chicago	452	New Orleans	139
New York	289	Louisville	132
Brooklyn	285	Buffalo	110
Boston	283	Minneapolis	101
St. Louis	275	Los Angeles	99
Baltimore	207	Detroit	94
San Francisco	205	Birmingham (Ala.)	92
Cleveland	192	St. Paul	90
Cincinnati	18(	Washington	85
Pittsburgh	16(		

One of the largest electric companies in this country published at the close of the year a list of the roads built by them, showing a total of 181 roads using 2,769 motor cars and having 2,264 miles of track; besides this they had 23 roads under contract.

At the beginning of 1891 the Thomson-Houston Company published the following list of electric railways, showing the relative distribution among the various electrical companies. Thomson-Houston roads, 103; Edison (Sprague), 83; United States Traction, 21; Van Depoele, 8; Rae, 8; Short, 8; storage, 4; Westinghouse, 3; Bentley-Knight, 1; storage, 1; total, 240.

A report from England states that there are now 29 miles of electric lines in operation in that country, on four different systems. Among the heavy electric lines at present in progress there are the Central London Railway (underground), 6 miles; and the Liverpool overhead railway, 6½ miles. A permit has been granted for the overhead trolley system on the Staffordshire lines, comprising 23 miles of track.

CHAPTER III.

CONSTRUCTION AND OPERATION.

The following opinions, information and advice, extracted and compiled from the papers of some of the best writers on the subject, being based chiefly on past experience, will give a very fair idea of the present practice in the construction and operation of electric roads. The matter is divided here into the following classes: Power plant, line, track, traction power; speeds, grades, loads, etc.; cars, miscellaneous.

POWER PLANT.

Mr. C. J. Field states: "The problem requires much more careful consideration than has been given steam power in electric lighting generally in the past. The work to be successfully done by the steam engine in the generation of electricity for the operation of railroads is of the severest kind, and can be compared only to that of the engine operating rolling mill trains. It is owing to not fully appreciating this fact that we hear in some parts of the country of failures of steam plants on this kind of work. Electrical manufacturers are assisting the solution of this problem by building the larger generators in units of 200 to 400 or 500 horse-power. What we want in the generating station for electricity is the smallest division of units consistent with the safe and economical operation of the station. Each unit should be entirely independent and separate from all other units, thereby increasing the reliability. This cannot be obtained in a safe and economical way by the use of the countershaft; in railway work, with large generators, we can see no excuse at the present time for its use. Generators should be belted directly to the engines, whether Corliss or high speed, or else coupled directly to the engine shaft. With a Corliss engine of 500 horse-power, operating at 80 or 90 turns, with a flywheel 18 to 20 feet in diameter, we can belt with belt centres of, say, 40 feet 2 inches, generators of several different commercial types; this gives us advantages which we have heretofore had only in high speed engines with direct connection. The engines should, in any event, as heretofore stated, be extra heavily built for the work to be done, with ample flywheel capacity. On engines of this size and speed a flywheel capacity of approximately 600,000 pounds is about right; on engines operating about 150 turns, say, 30,000 to 40,000 pounds. Generators on this work are subjected to the severest and most excessive strain, particularly where of small type, but the building of them in larger units is going to remove, to a great extent, the question of the overloading of the machine. Railway machines are often subjected to overloads of from 25 to 50 per cent. In general these are only momentary, and we find most of them able to stand up to the work to be done.

"High speed engines in the development of railway work have received in some cases a setback, owing to the engine manufacturers not appreciating fully the conditions and necessity of the work undertaken. So called high speed, or automatic engines, can be as successfully operated on this class of work as any other, if they are especially built for it. This means larger parts, bearings of more ample size and length, and ample flywheel capacity. On a cross compound engine of, say, 300 horse-power, there should be about six to eight tons in the flywheels, the bearings seven or eight inches in diameter, and 15 or 18 inches in length. In the case of engines built in this manner there can be no fault found with their operation. A type of engine, which we believe is going to be largely used on this class of work, as well as lighting work, is one that will come in between the high speed engine and the Corliss and which will combine many of the advantages of both. Such an engine has been sought for by many engineers, and has been attempted by a number of builders. To-day, however, we cannot find it on the commercial market. This engine, in units of 500 horse-power, would run at a rotative speed of about 140 or 150 revolutions, and with a piston speed of about 650 to 700.

"The question which has troubled most engine men in regard to the high speed engine with a single valve, covering this kind of practice, has been a question of valves and clearances. Beyond any question, when it comes to this size, we have got to come to the Corliss practice of double valve, thereby reducing the clearances and bringing it down to the extent of the Corliss practice. The trouble in this line has been to get electric manufacturing companies to take up the building of large multipolar generators adapted for direct coupling at a speed of from 100 to 200 revolutions. This problem was developed on a much smaller scale in this

country, for marine plants, several years ago. We
find that in Europe, where their work has been
more special, that they have successfully developed
this type of engine and generator, and, beyond any
question, it is going to be both for lighting work
and for railway work the type of unit for central
station practice in the future.   It means, where the
vertical engine is used, the installation of the steam
and electric plant in the space formerly used for
engines alone.   This means reduction in the cost of
building, operation, and maintenance.

"We append a few interesting figures and data
which the writer collected for presenting to street
railway companies, in order to give them some
useful information in this respect. The figures given
in the table, etc., are not ones that the manufacturer of an engine would tell you were those of the
best economy for his engine or plant, but they are
figures which will be appreciated by station owners
and railway companies as those which are obtained
in every day commercial tests.

"The relative commercial economy of engines and
cost are as follows:

TYPE.	Lbs. of coal per h. p. hour.	Cost per h. p., sizes over 100 h. p.*
High speed single..................................	4 to 5	$11 to $13
" compound.............................	3 to 3½	14 to 16
" com. cond............................	2¼ to 2½	
" com. triple...........................	1¾ to 2	18 to 22
Corliss single....................................	3½ to 4	16 to 18
" compound cond.......................	1¾ to 2	22 to 25
" triple................................	1½ to 1¾	27 to 30

* This is based on an evaporation of 9 lbs. of water per pound of coal.

"There are four classes of boilers:

"1. Horizontal return tubular, whichis the most general in use, and costs $9 to $10 per horse-power.

"2. Vertical tubular (Corliss or Manning), which is a vertical tubular boiler, with water leg, giving an internal fire-box, economical in floor space, largely used throughout New England. Cost $10 to $12 per horse-power.

"3. Sectional or water tube boiler, of which the Babcock & Wilcox is the best known, especially adapted for higher pressures and safety. Cost $17 to $19 per horse-power.

"4. Scotch type of marine boiler—one that has not been used to any extent as yet in station work—but we believe it will be as an offset to the sectional type, and fulfilling the requirements for higher pressure and economy of space.

"The capacity of engines requisite for different generators at a steam pressure of 100 pounds:

GENER-ATOR.		Engine.					
		High Speed.			Corliss.		
Watts.	Horse-power.	Size.	Speed.	Wt. 2 fly-wheels.	Size.	Speed.	Wt. 2 fly-wheels.
50,000	75	12 × 12	280	7,000 lbs.	....	....	....
80,000	125	18 × 16	225	9,000 lbs.	....	....	....
150,000	225	18½× 18	200	15,000 lbs.	20×36	90	25,000 lbs.
2,150,000	450	....	....	....	24×48	80	50,000 lbs.

"The cost of steam plant complete is about $50 to $60 per horse-power for high speed, and $65 to $75 per horse-power for Corliss.

"We desire to call the attention of central station

owners to the profit to be made from the furnishing
of power in street railway operation, and also by
the combining in smaller towns of the street railway companies and electric light companies. The
trouble in most cases in central stations obtaining
;contracts for power, outside of small roads, has
been to convince the railway companies that the
electric light station can economically and reliably
furnish this power, and we must say that in many
cases their fears are well founded. Therefore, it
behooves the central station companies to place
their generating plants and stations, not only for
their own business, but for this added business, in
such a shape as to remove this objection. There is
no reason why electric light stations should not do
a large and profitable business in this line as well
as in stationary motor work, for the same factor is
introduced here, and the same reasons why they
can safely and profitably furnish this power; if
they have a station properly built, and large enough
to add this power, that factor is established. If they
have a proper station operating force, in many cases
this force need not be added to at all. As to what
basis this work can be probably done on, we hesitate to state figures, except in specific cases, but
will try to give a general idea of some of them. For
many small roads power contracts have been taken
at so much per day, assuming a basis of 100 to 125
miles operated. Such contracts have been at from
$3 to $5 per car. The regular basis, in accordance
with which most street railway companies make
their contracts and desire to base their cost of
operation, is the unit of car mile operated; there-

fore, most contracts are on this basis. This comes down, therefore, to a basis of from three to five cents per car mile; the latter figure we consider excessive, and one which would be only made by any company for temporary necessities. We know of cases where the matter has been carefully considered and the plant properly installed for it, where contracts have been made for between 2½ and 2¾ cents per car mile for 16-foot cars, on roads with grades not exceeding 1½ to 2 per cent. In this case, and, in fact, in most cases where the closer figures prevail, the railway company furnishes the generators and the station owner furnishes the steam power and all expenses of both steam and electric power due to ordinary wear and tear. A profitable source of investment has been found in the more moderate sized towns of, say, up to 30,000 or 40,000 inhabitants, in the installation of combined electric railway and lighting stations; the companies either equipping new ones or purchasing old street railway systems and dilapidated lighting plants running on an unproductive basis, but which have a good franchise and field for business. Such companies have proved very profitable, as the combining of the operating expenses for railway and lighting station has done much to reduce expenses, and in many cases one manager or superintendent has proved sufficient for the entire system."

Mr. Beckley advises not to make the units in a power station too large: "Accidents will happen as long as machinery is run, and an accident to a 500 horse-power plant is serious, while you can keep your cars or most of them moving if

one of two or three small engines breaks down.
The same rule, of course, holds as to the generators. Always put in a condensing steam plant.
One large item of expense of operation is the
coal bill. Cut that down at least 40 per cent. by
erecting condensing engines. The first cost is, of
course, a little more. Locate your power station as
near as may be in the centre of your system, but,
above all, if possible, on a stream large enough to
furnish all the water you require for the boilers and
condensers. City water, where your consumption
runs into the millions of gallons, is expensive."

The following test of the power plant of the Utica
(N. Y.) Electrical Belt Line Railway, made by
Messrs. Heilman and Clarke, contains some figures
and results which may be of interest and use, as the
test was probably carefully made. It is to be regret-
ted very much that such a good opportunity was
lost to note other data also, such as the number of
cars, etc., from which very useful results could have
been deduced. As it is, it is simply a boiler and
engine test, but as such, it is quite interesting. This
company operates 26 miles of road. Twenty motor
cars, and a number of "trailers," used when traffic
is heavy, comprise the rolling stock of the system.
The steam part of the plant consists of three 200
horse-power Armington & Sims cross compound
engines. Only two of· the three engines are constantly in use during the day, the third being
required only at intervals, when heavy loads are
suddenly applied. Its use was not required at any
time during the test. The boiler room contains four
horizontal tubular Curtis boilers. These boilers are

each 6 feet diameter of shell, 16 feet long, and contain eighty-eight 3½ inch tubes. Two boilers only are used at a time, a change to the other set being made when cleaning or repairing is necessary. Two independent boiler feed pumps and one independent Davidson air and circulating pump with jet condenser are provided. A National feed water heater is placer between the boiler and feed supply, while a separator was placed on the main steam pipe, being furnished with a Curtis regulator to return the separator water to the boiler. In the generation of electricity, six Thomson-Houston dynamos are in use, two being driven from each engine. The instruments used for measuring currents and voltage are also made by the same company.

Two runs were made, a preliminary run on May 2d of five hours' duration, and the main test on the following day of ten hours, readings being taken from 9 A. M. until 7 P. M. The feed water was measured by means of two calibrated barrels, one being filled while the other was emptying, and then reversing the process. Although the quantity of water required was large, by the use of this method no difficulty was experienced in keeping the supply equal to the demand. The injection water for the condenser was drawn from a pond near by, and after passing through the system was again returned to the pond through a sluice of rectangular section. The quantity of flow was obtained by placing a weir in the sluice, and the height of flow over the weir was measured by a hook gauge. The weight of coal consumed was determined in the usual manner, a Fairbanks scale being placed in the boiler room, and loads of 200 pounds each were dumped in front of the boiler at a time. The ash-pits were cleared at the end of the test, and the ashes placed in a barrow and weighed. In order to obtain the quality of steam at the engines, a calorimeter was connected to the pipe leading to the oiling apparatus, the connection being made close to the engine throttle. A comparison of results taken from boiler calorimeter and engine calorimeter shows a wide difference in moisture in steam. Electrical readings of current and voltage were taken every five minutes during the test. Below are tabulated some of the more important results:

Character of draught...Natural
Temperature of external air..............................54.5 degrees Fahr.
Temperature of boiler-room...68.9 degrees
Duration of test..10 hours
Fuel burned...7,000 pounds
Total ash..988   "
Moisture in 100 pounds...8   "
Weight of combustible...5,472   "
Efficiency of combustion...82.15 per cent.
Kind of coal—One-third bituminous slack and two-thirds anthracite culm
Fuel per square foot of grate per hour...........................11.66 pounds
Combustible per square foot grate per hour.....................9.087   "
Per cent. entrained water (boiler cal.)...................1.24 per cent
Total water supplied...49,355.4 pounds
Dry steam per hour from temperature feed water.............4,875.1 pounds
   "   "   "   "   " 212 degrees...........................5,414.22   "
   "   "   " pound of coal.......................................6.96   "
   "   "   " square foot of grate.........................81.25   "
   "   "   "   " " heating surface...................1.67   "
Thermal units per hour...................... { given   7,805,400
                                            { received 5,793,978
Horse-power (34.5 pounds water evap. per one horse-power per hour).......143
Efficiency of boiler...73.29 per cent.
Average steam pressure (absolute)...........................133.8 pounds
Temperature of feed water to heater.........................77.2 degrees
   "   "   "   " from heater...........................164.5 degrees
Change of temperature due to heater.......................87.3 degrees
Revolutions of engine...261
Per cent. variation of speed....................................1.14 per cent.
Average indicated horse-power....................................190.53
Quality of steam at engine.................................3.88 per cent. wt.
Coal per indicated horse-power per hour............................3.5 pounds
Water per indicated horse-power per hour (without charging
   condenser)...21.72 pounds

$$\text{Average efficiency} = \frac{\text{Average electrical horse-power}}{\text{Average indicator horse-power}} = 81 \text{ per cent.}$$

LINE.

Regarding the usual practice on various overhead trolley roads, and other like data, some very interesting information was compiled by Mr. Mansfield from a large number of circulars sent out by him to all the various roads in operation, asking certain questions. He received answers from 137 roads, operating 1,546 miles of trolley wire and 1,657 motor cars. Of these 71 were Thomson-Houston roads, and 66 Edison or miscellaneous. He summarizes the answers as follows:

"The trolley wire: Of the 137 roads 99 were using copper wire, 28 silicon bronze wire, eight were using both, and two were using phosphor bronze.

"Of those using copper, one used No. 000; five No. 00; three, No. 1; two, No. 2; one, No. 5 hard drawn copper; one used No. 0 soft drawn copper, making 13 in all, and leaving 86 as the number using No. 0 hard drawn copper wire.

"Of the 28 using silicon bronze, 16 use No. 4; six, No. 2; one, No. 3; and the five remaining roads had combination, of two or more sizes.

"Of the eight using both silicon bronze and hard drawn copper, six prefer the latter.

"Out of all these 99 using copper, not one dissents, but of the 28 using silicon bronze 11 advise copper. The proof is conclusively in favor of hard drawn copper wire and of the larger sizes, No. 0 B. & S. seeming to be the standard.

"In regard to the wearing, the universal testimony is that it is exceedingly slight. What wearing is observable is found to be at the switches or on the curves.

"Serious mistakes have been made in the past by using iron flanged trolley wheels. These cut the trolley wire badly. Everything should be done to throw all the wear on to the trolley wheels. These are much less expensive than wire, and not so hazardous for the public if they give out.

"From my own information and knowledge, I can say that the life of the trolley wire is much longer than I originally anticipated. The criterion is not the years that it has been up, but the number of times the trolley wheels have passed over it. It would seem that with ordinary brass trolley wheels the wear was about .001 of an inch to the passage of 65,000 cars. This is at the rate of one in every six minutes, for 18 hours per day, for one year. With only this wear the life of the wire would certainly be 20 years, unless through some process of crystallization it became more brittle. I anticipate but very little trouble in this direction, but eternal vigilance is nevertheless necessary. Undoubtedly, at curves and on switches the wear is somewhat greater. How much I cannot say.

"The breaking of the trolley wire has been rare, the breaks occurring either at splices or switches, or were due to some extraneous cause, such as falling trees, telephone poles, or the catching of the trolley pole. One road reports a break as due to the striking of the trolley wire by a locomotive smokestack. In no instance was any casualty reported, excepting in one case, where a mule was killed.

"Forty-one roads have their trolley wires divided into sections, and consider it necessary and advisable. Analysis shows that these roads are in the largest cities or towns.

"As to the relative wear of the sliding and the rolling contact trolley I believe only one road used a sliding contact trolley, and I think that has been abandoned.

"Span Wires.—Regarding span wires, forty-nine report as using galvanized iron wire, fifty-five as using galvanized steel, twenty as using galvanized iron cable, and one as using copper wire. The sizes range from No. 0 to No. 14. Comparatively few breaks are reported and no casualties.

"My own experience has led me to adopt No. 4 B. & S. soft galvanized iron wire. Whenever a long span or a curve is to be constructed I have had two of these wires twisted together into a cable.

"I have found that a cable made of small wires is hard to joint, and it rusts much more quickly. Avoid joints, and use a ball fastener in attaching span wire to eyebolt. On the whole, however, I have concluded that iron is not the proper material to use in any shape. It will rust, and then your structure is weak. Pursuing my investigations into this matter a year ago, I found that a certain special quality of silicon bronzed wire was the best. Tests of this wire, in comparison to iron, showed the following results:

	Diam.	Breaking weight.	Breaking weight per sq. in.	Elongation in six feet.	Twists in six feet.
No. 1. silicon bronze	.200	2.550	81,800	.8 per ct.	37.4
Galvanized iron	.205	1,720	42,000	7.8 per ct.	19

"I am aware that the price of this wire is five or six times as great as that of iron wire, but as the total sum in either case per mile is small, I strongly recommend it. Some of the wire has been in service on the West End road in Boston for nearly a year. It certainly will never rust out.

"Guard Wires.—Guard wires are universally condemned, but are put up as a compulsory protection from existing evils. When there is only one trolley wire to protect, we extend two light wires, about No. 14, on each side of the trolley wire, and about 18 inches above it and from 12 to 18 inches to each side. We suspend the longitudinal wires running parallel with the trolley wire upon an additional span wire. The additional span wire is insulated as perfectly as possible from the poles and from the trolley span wire, and the longitudinal or guard wires proper are also insulated as far as possible from the span wire. In the case of two trolley wires, the general practice, so far as 1 have observed, is to stretch three guard wires, two of them outside the two trolley wires and one over the center, all insulated perfectly from the other wires. I think they are of great value. I think it necessary to put them up. Most municipalities require them to be put up, and as long as you keep them perfectly insulated from the trolley wires, any extraneous wire which may fall will strike them, instead of the trolley wire, and will dangle in the street as a dead wire, which may be removed.

"Feeders.—The descriptions of the various feeder systems are so vague I will not attempt to describe them. The average distance to which power is transmitted on these roads is about three miles. The greatest is 10.7 miles on the Tacoma & Steilacoom

Railway, Tacoma, Wash. There are many, however, operating from eight to ten miles from the station.

"There are two general methods of operating the overhead system. One is by having a continuous trolley wire wherever the track runs, and the second is to have this trolley wire divided into sections. For towns and for suburban traffic the former is almost invariably adopted and carried out. Practice would seem to indicate that but little trouble is experienced, and that there is practically no advantage in dividing the trolley wire into sections, in fact, a disadvantage, since you lose its conductive capacity. The sectional trolley wire surely must be used for all city work. It would be practically impossible to operate in Boston without proper divisions. It is, however, almost impossible to originally fix all of these divisions once for all, as the topography of the city, the grades, location of power houses, routes, lines of traffic, etc., all have to be taken into consideration. It is bound to be a gradual growth to a large extent.

"Obviously the methods of feeding the trolley wire vary with the methods of arranging the trolley wire. With the first method mentioned (using a continuous trolley wire) the feeders are either extended from the station the entire length of the line, tapping into the line at intervals, or else separate feeders are run out from the station to certain predetermined distances, and there tapped into the trolley wire. When more than one feeder wire is needed, in either case a repetition of the scheme is carried out from feeder wire to feeder wire; there is little to choose between the two methods; both are good.

"With the second method (the divided trolley wire) there are two ways of accomplishing the feeding. First, to extend a feeder the entire length of the line and tap into the centre of each section of trolley wire; or second, to extend the feeder the entire length of the line and tap into both ends of each section. The advantages of the former are that in time of trouble a man has to run to only one box to cut out a section, or the whole arrangement could be made automatic by putting a fuse, or mechanical circuit breaker, in the box. The disadvantage is that you lose the value of the trolley wire as a conducting medium, which in the case of hard drawn copper wire is considerable.

"In the latter method (tapping into the section at each end) obviously a man has to go to each end of the section to entirely cut out that section, and two sets of automatic devices would have to be arranged to operate in case of trouble. You have, however, the advantage of utilizing the trolley wire as a conductor.

"It is undeniably true that the question of feeder wires is one of great importance and a difficult one to always economically solve for all conditions. The point, however, which the railroad corporations should watch above all others is that they have enough. I have visited many roads where I found that the larger part of the trouble which they were complaining of lay in the fact that they did not have either sufficient trolley wire or track feeders. Railroad officials are very apt to object to feeders, because of, first, their cost, and second, the placing of so many wires overhead, which is liable to bring them in conflict with the municipalities and the public. The question of having an insufficient number of feeders underground for the track is one which they can have no valid or reasonable excuse for.

"I am strongly of the opinion that for large cities all feeders should be placed underground. The cities in which this underground work has been adopted are Buffalo, Minneapolis and St. Paul. The question is one almost entirely of construction. The cables alone will cost in all probability less than overhead wires. This construction work can be done simultaneously with the track reconstruction, for it is my experience that whenever a large city railway adopts electric power it is almost absolutely necessary to rebuild its tracks. Under these circumstances 1 doubt if the cost of the conduits or ducts would be more that a few thousand dollars additional per mile.

"Ground Plates.—I strongly recommend as many ground plates as it is possible to have, not only at the station, but also along the line; brooks, bog land, water pipes, everything should be utilized for this ground circuit wherever possible. The plates can be made of sheet copper or iron, preferably the latter, and should have a superficial area of several hundred square feet. The wire connecting them to the tracks should be of sufficient size and very solidly attached to the plate and the rails. The continuous supplementary wire should in all instances be employed, and the rails bonded at least once. In no instances do I think it necessary or wise to place the track feeders overhead.

"Accessory Devices.—In regard to the overhead devices and material used, I can only urge the advice that the most substantial and perfect apparatus that can be secured be used. Too much care and attention cannot be bestowed upon these devices. It does not pay to put in some little cheap fifteen-cent arrangement, when for fifty or sixty cents a substantial, reliable, and standard device can be obtained. It is also well to consider the question of uniformity in the apparatus. The only part that is liable to deterioration is the insulating material. Make this, therefore, of a uniform pattern, and arrange the various holders for its reception."

Regarding overhead line construction, Mr. C. J. Field states: "We find in the past about as great a development in overhead and line construction for electric work as in any other part of the subject. While formerly this was one of the greatest sources of unreliability in the operation of the plants, to-day it has reached a very practical development. In the insulation of a single trolley system, with one side of the system grounded, we have the most severe requirements that it is possible to obtain in any electric insulation, in that any grounding on the other side of the system means trouble in operation of the road. This led to the introduction of double and even triple insulation into our line material to properly protect the trolley wire from grounding. Where streets are wide enough to spread the tracks to six feet, and six feet six inches within the near rails, we see introduced in many places centre iron poles, which make a considerably stronger style of construction than cross suspension. There are not many streets, however, where street cars are in operation that are wide enough, or where the city will allow the spreading of the tracks to this distance, and in closer proximity it is not safe to operate with centre poles."

Mr. Beckley states: "The overhead wire cannot be too well put up. Cheap devices should never be used because they are cheap. The best and strongest are none too good. In putting up the feed wire and putting in the ground wire return to the generators, do not spare copper. I am convinced that much that we have heard about the inefficiency of generators and motors is due to trying to get too great a quantity of current through too small a quantity of copper."

Regarding guard wires, Mr. Bickford states that nothing short of No. 10 should be used. The idea is not so much as to its life as its additional strength in supporting fallen wires, especially when loaded with sleet in winter. On a portion of their lines they had broken telephone and telegraph wires on the guard wires for a long distance, and through it all they operated their cars without interruption by using a No. 10 steel wire.

Mr. Littel thinks it is advisable to cut up the guard wire into sections, of say 1,000 or 1,500 feet, and put in circuit breakers, so that if any wire does fall, the current will not be carried a great distance. It is done on many roads. The West End Road, in Boston, is said to do so with the line every 500 feet; it is done in Buffalo also.

According to the experience of Mr. Wason, galvanized iron span wires used in Cleveland were
rusted so badly at the end of a year that they did
not feel safe to tighten them.   They claim to have
been the first to use No. 4 soft drawn copper for
span wires, which has been up for a year and there
has been no perceptible elongation or stretching.

Regarding the trolley wheel and poles, Mr. Everett advocated having a wheel that is capable of following the wire at any angle, with a trolley pole
brittle enough to break should it become entangled
in the wires, without pulling them down, and a
trolley pole rigid enough to give good, steady pressure on trolley wire, and so constructed that when
the car is in the car house or going under a low
bridge the pole could come very close to the roof of
the car, also flexible enough to give good pressure
when the trolley has to be 21 or 22 feet high at the
railway crossings.

Among the articles published was a short serial
by Mr. Arthur E. Colgate, on "The Construction
and Care of Electric Railroads."   Being eminently
practical in character, and containing many useful
hints, we have reprinted here in full that portion
which refers to the line: "The power which may be
said to be lost in the resistance of the trolley and
feed wires is easily calculated by the formula,

$$\frac{C^2 R}{746} = \text{the horse-power lost.}$$

"The resistance of a copper wire is very nearly
what is expressed by the formula 32.37 times its
length in yards, divided by the square of its diam-

·eter in thousandths of an inch; and for other metals the resistance may be found in the same manner, the constant being changed according to the relative conductivity compared to that of copper. For example, if the conductivity is half that of copper, the constant should be multiplied by two. The formula is based on the resistance of a copper wire one yard long and of one circular mil sectional area, and as the conductivity of silicon bronze and most other trolley wires is about 25 to 40 per cent. higher than that of copper, either the constant or the result should be multiplied by 1 plus the per cent. increase of resistance.

"Where feed wires are run in connection with the trolley lines, and are cross connected with them at short distances, the joint resistance may be considered as that of one conductor, whose resistance is equal to the product of the several resistances divided by their sum, and the loss of power calculated in the same manner as it would be in one wire. The loss permissible on a line depends greatly on the grade and cost of power. In most cases, where coal is reasonable, it is about 25 per cent. of the total amount developed by the generators, and on grades it should be much less, the current being supplied at these points by a separate feed wire, and a circuit breaker being put in both trolley lines at the top and bottom of the hill, if it is a short one; and if it is sufficiently long to warrant it, the circuit should be again broken in the centre of the grade. The object of this is to protect the rest of the system, should any car motor burn out or the line become grounded through defective car wiring, the insulation of which would be liable to give way and ground the line when subjected to heavy service on grades.

"The weight of copper wire is about equal to the square of its diameter in thousandths of an inch multiplied by the constant (which is .016), and its tensile strength 11 times its radius in thousandths of an inch squared multiplied by the strength of a square inch of the material, which, in the case of copper, is 61,200 pounds, and in the case of iron 103,-000 pounds. The strength of cable wire is about 10 per cent. more than that of the various strands of which it is composed, and the strength of rope is expressed by the circumference squared multiplied by the constant, which is for white rope 1,140 and for manila 810, the strength of tarred ropes being about 10 per cent. less. The weight of round iron poles is 2.65 times the mean outside diameter multiplied by the strength of the pole.

"Where the use of untried apparatus is contemplated, tests as to the relative strength and durability under strain should be made by means of a testing machine. Imagine a case in which a log of wood about ten feet long and eight inches in diameter is held in an upright position by suitable means. Projecting pieces of wood are bolted to the log by inch bolts or lag screws, there being two chains of half-inch iron links, kept from sliding together by means of projecting bolts. A small Fairbanks dynamometer is used, but, if not procurable, may be replaced by a steelyard. There are two clamps, one of which is attached to the dynamometer and the other to the turnbuckle by means of a swivel. The piece of apparatus to be tested, having had short pieces of suitable trolley wire attached to it in the regular manner, is held by means of these and the clamps, and the strain applied by the turnbuckle, the amount of which should be read on the dial of the dynamometer. Should a steelyard be used, it will be much more convenient if a small spring balance is substituted for the weight or counterbalance.

FIG. 2.—TESTING A POLE.

" If it is thought necessary or desirable to test any of the various makes of poles, it may be done as is shown in Fig. 2, in which strong pieces of wood are firmly set in the ground, and a stout piece of hard wood is placed behind the pole to distribute the strain evenly along that part which is afterward set in the ground. The strain is applied and read in the same manner as in the previous case, by means of the dynamometer and turnbuckle. In applying this test to side and pull-off poles, the rope or chain by which the strain is brought to bear should be attached to the insulated cap at the top, which is usually the weakest part of the pole. This has proved so weak in many instances that constructors are discarding it, and are attaching the span and

pull-off wires to circuit breakers, which are attached
to the pole a few inches from the top by a short
pendant and turnbuckle, or by means of an insulated turnbuckle. These methods, while not presenting so symmetrical an appearance, have a most
decided advantage of superior strength and durability.

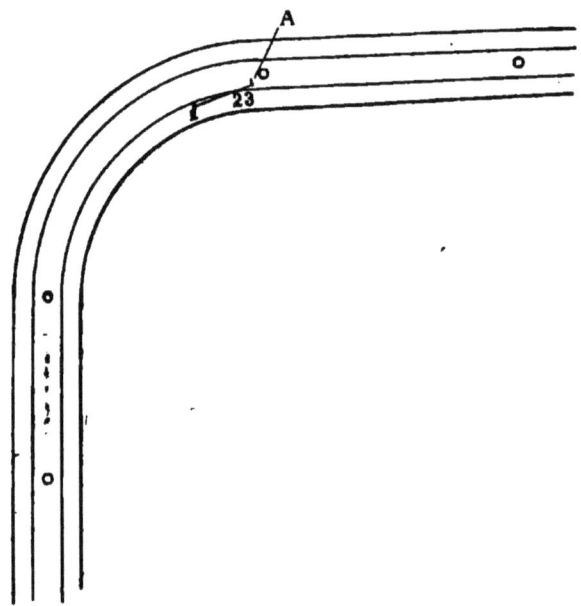

FIG. 3.—LOCATING A CURVE.

"In centre construction a pole should be set as
near each end of the curve as possible without
involving danger of being struck by the overhang-
ing ends of the cars. The precise point for the location may be found as is shown in Fig. 3. A line with
the distances corresponding to the length of the
wheel base and distance from the rear axle to the
front dashboard of the car marked on it is held

between three men. The men who represent the wheel base, 1 and 2, stand on the inside rail, the other standing in line with them at the point representing the forward dasher. A short stick is held by him at right angles to the line. The curve is followed in this manner, and the location should be fixed at a point where the stick clears a line drawn

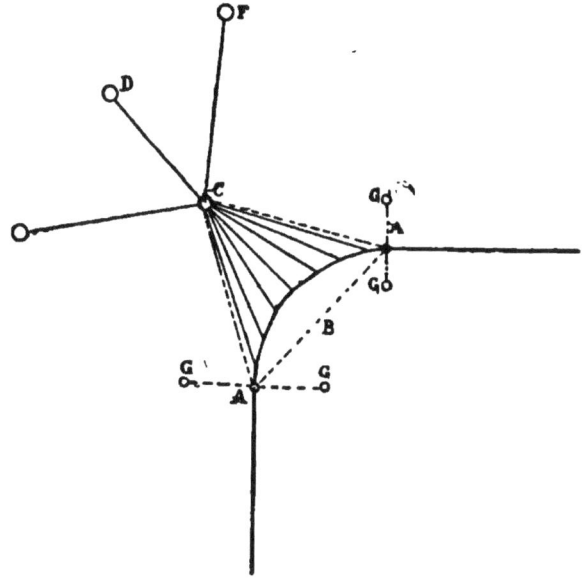

FIG. 4.—LOCATING A PULL-OFF POLE.

parallel to the rails between the tracks by a little more than half the diameter of the pole.

"Fig. 4 shows the correct location of a pull-off pole, which is found by an equilateral triangle B, having for its base a line drawn between the last two poles, located at the ends of the curve A A, the apex of which is the pull-off pole C. If it is not convenient to set it here, it may be moved farther back, or a short distance to the right or left, as circumstances may require, or the pull-offs may be brought together at this point and made up into an iron ring with a pendant running to the pole, or two poles may be placed at the points E and F, and the ring secured by pendants running from them.   In the case of side pole and span wire, construction poles should be placed at the points represented and a span wire run between them, as shown by the dotted lines, the object being to support the curve and

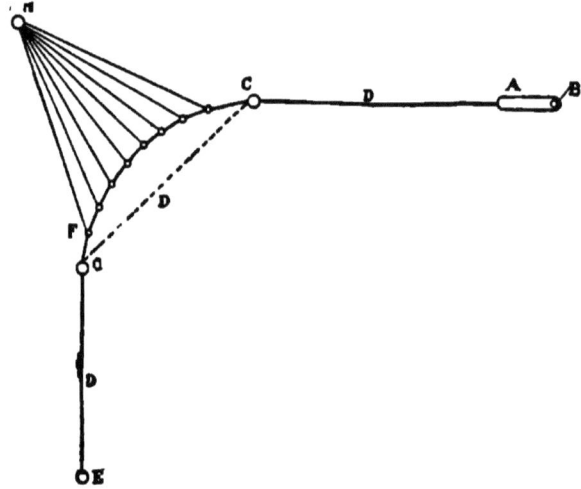

FIG. 5.—MODEL FOR POLE LOCATION.

keep slack from running back should the pull-off pole bend to any extent, or should the wire part. The ears by which the trolley is fastened to these spans should in all cases be soldered very firmly with a solder composed of equal parts of tin and lead soldering, salts being used as a flux.

"Poles are often located by means of pins driven into a card on to which a sketch of the curve and line has been pasted, the line being in this case represented by a thread which is pulled out over the track in the same manner as the trolley line should be. In the diagram, Fig. 5, A is an elastic band fastened to the board by a nail B, C C being the pins which represent the poles at each end of the curve, around which the thread is run and anchored to the pin E. F F F are pins driven into the board on the curve line at a distance of eight or

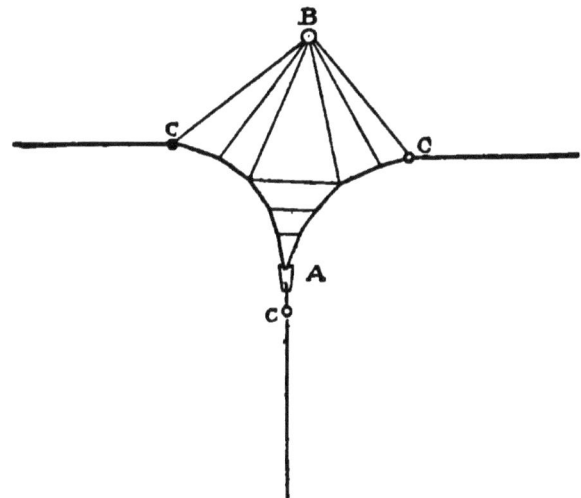

FIG. 6.—PULL-OVER FOR DOUBLE CURVE.

ten feet apart, on the scale to which the sketch is drawn, which serves as a guide to the length of the pull-offs, and which are attached to the pull-off pole H, on the curb line. After the line is pulled out against the pins they are removed, and if the thread follows the curve the location of the poles is correct.

"When there is a switch in the line, where it branches in the shape of a Y, the curves may be pulled from each other as is shown in Fig. 6, in

which A is the switch, B is a pull-off pole, which is
located in line with the switch A, and C C C are the
last poles on the straight line. The switch A should
be situated about four feet beyond the switch in the
track. Turn-out switches should be located in the
same manner.

"This chapter would not be by any means complete

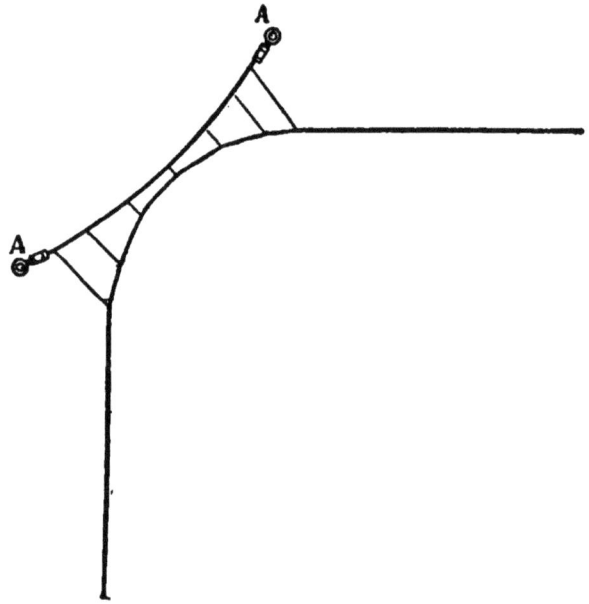

FIG. 7.—SHORT'S ARRANGEMENT OF PULL-OVERS.

without speaking of the most excellent manner of
curve building devised by Mr. Sidney S. Short, of
the Short Electric Company. Two poles A, Fig. 7,
are set on opposite curves, and a heavy span wire
run between them, and insulated from the poles by
circuit breakers. Short pull-offs are then run to the
trolley wire from the span wire.

"All poles for electric railroad lines should be set

either in concrete made of Portland or Rosendale cement, sand and broken stone; or, in case of bracket or centre pole work, they may be set in cobble-stones and earth. The earth in the latter

FIG. 8.—WAGON FOR POLE ERECTION.

case should be well tamped by means of wooden or iron tamps.

"One of the best ways of setting the poles, after the holes are dug, is by the use of a tower wagon, which is made according to Fig. 8. To the body of a wagon are fastened four uprights, the tops of which should reach 19 feet above the ground; and these are stiffened by four braces, which are made of hemlock about six inches wide and two inches thick.

"Across the two front uprights run boards, which form a ladder, and this enables a man to ascend to the platform, which is about 7 feet × 4 feet 6 inches, and has a slight bulwark around it one foot high and lockers on each side for tools. At 12 feet from the ground is the arm, which is securely fastened at two points. A winch is fastened to the body of the wagon, a little back of the centre. A rope leading from this is led through a snatch block, and is fastened to the pole a little above the centre of gravity.

"By this means a pole may be set in a very short time by three men. After the pole is in the proper position and the proper rake given it by means of a spirit level, which should have a taper on one edge, it is held in place by pike poles and the cement packed in around it. This concrete should be made in a large dump cart, which can be driven from one hole to the other, and the cement shoveled directly into the holes, and well tamped in around the poles.

"About a week after the poles are set the arms should be attached; and this may be done by means of the tower wagon or by men using ladders, depending largely on the nature of the arm used.

"The ladder should be five feet longer than the pole and should be lashed against it top and bottom. A snatch block is attached to the upper rung and a rope run through it, by means of which the arm is raised, one man being at the top of the pole to guide it into place and fasten it there with bolts. All pins, etc., should be attached to the arm before it is put in place.

"Erection of Feed and Trolley Wires.—After the poles are set and properly guyed, the feed and trolley wires should be run, starting from circuit breakers which are firmly anchored to heavy poles, either on the side or between the tracks. These poles should be set very firmly in concrete to a depth of seven feet and have a rake of about ten inches. The feed wire should be spliced by means of soldered Western Union joints into lengths of about a mile to a mile and a half each, and wound on a large reel firmly mounted on a flat truck, sufficiently large to allow two or three men room to work comfortably. The end of the coil is attached to the circuit breaker and about 60 feet run out, which is kept clear of the ground by passing it over a drum mounted on the tower used in setting poles. This follows the truck bearing the reel. The wire is kept sufficiently tight to prevent much sag between the poles by a wooden bar which is held against the side of the reel by two men. After a certain amount, depending on the size of wire, has been run out, it will be found difficult to keep it sufficiently tight to keep the slack clear of the ground. At this point a wire head guy should be run from the top of one pole to within ten feet of the base of the next. The feed is pulled tight by means of blocks and fall attached to the top of the pole and to the wire by a small chain or rope strap, which should be braided around it. After being pulled tight the wire is made fast to a pole back of the blocks, which has been previously guyed in the same manner, and the blocks and fall removed.

"The line should then be tied in or fastened to the insulators with short tie pieces of wire by linemen, who go back over the line for this purpose, the tower and reel wagon keeping right ahead, paying out and hanging up wire. At the next pull up one man stays back and unfastens the strap which holds the line after it is pulled sufficiently tight to prevent slack from running back into it through the ties.

"Just before turning curves the line should be pulled tight and secured by straps, and should be again pulled up at the other end of the curve, but not quite to such a degree of tightness as the rest of the line. Where it is desirable to drop one feeder it should be secured to a circuit breaker which is fastened to a pole by means of a pendant and a turnbuckle or by an insulated turnbuckle. The largest feed wires are run before the smaller ones and on the pins nearest to the poles.

"If it is thought necessary to run guard wires they should be made of steel not smaller than No. 5 B. & S. gauge, which in cities should be painted with waterproof paint or Nubian enamel, as the zinc coating is soon destroyed by the chemical action due to the smoke in the atmosphere. They are run very much in the same manner as the feed wires, about two feet above the trolley lines, and pulled very tight. These guard wires should have circuit breakers in them at regular distances of about one thousand feet or less, which may have a lead wire or jumper around them if the wire is to be used for signaling purposes.

"In the case of side pole construction, where the feed wires are run on arms attached to the span

FIG. 9.—METHOD OF CROSSING LINE WITH FEED WIRES.

poles, the feed wires where they cross should be raised above the trolley and guard lines by means of raised arms similar to the one shown in Fig. 9, which is clamped to the pole by means of iron

bands and is kept from turning by set screws. The
span wires, with the hangers attached, should now
be put in place, after the completion of the feed
lines, and the trolley wire run in a similar manner,
the only difference being on the curves and joints
and in its being hung under the arms or span wires
by means of hooks made of heavy steel wire. There
is also a great difference in the manner of making
splices, which is by the use of splicing ears that are
put in the line after it is pulled up tight, the wire
being held temporarily by clamps and turnbuckles,

FIG. 10.—METHOD OF HOLDING LINE PREPARATORY TO MAKING SPLICE.

as shown in Fig. 10. These ears should be attached
to the arms or span wires in the same manner as
the regular supporting ears. The practice of making
Western Union or sleeve joints in these lines is, I
hope, a relic of the past. After a reasonable amount
of wire, say a mile or so, is run out, or a curve is
reached, the wire is drawn up sufficiently tight to
allow about eight or nine inches sag in summer and
about five inches in winter, depending largely on
the size and strength of the wire used, and it is held
in place by a head guy clamped to it running to the
base of a pole. The turnbuckle at the joint is then
drawn up about a foot, and the exact location of the
splicing ear marked on the wire with chalk or a
burnisher, but under no circumstances should it be
scratched there with a file. The surplus ends are
then cut off with a bolt cutter, which is, by the

way, the only thing which will cut trolley and cabled wires and make a good job of it. The splicing ear is then put in position and the wire fastened to it in the regular way according to its kind, when the turnbuckles are slacked off and the clamp removed. Circuit breakers are put into a line in an exactly similar manner.

"After the splicing ears and circuit breakers are in place, the section may be fastened to the supporting insulators either by soldered ears or good mechanical clips. If the former are used they should be soldered by means of a copper bolt, and not by either pouring the solder on to them or by the use of blow lamps, the former being weak and the latter injuring the wire. And under all circumstances soldering salts should be used instead of acid or resin. On curves the trolley wire is pulled out over the track by loosely attached temporary pull-off wires running to the poles, their location having been ascertained, as described in previous pages. Their length is just sufficient to allow the trolley to curve about one foot to the inside of the centre of the track, its exact location being determined by the radius of the curve and the nature of the trolley stand, together with the judgment of the constructor. The outside wire is first run, and the inside one pulled from it by a continuation of the pull-off wires.

"Line switches should be so constructed as to admit of their being moved if necessary, and should be without moving parts. All movable apparatus intended to be operated by the trolleys passing under it and by gravity is liable to become clogged with grease or ice, and is, as a usual thing, either weak or bulky. Where these switches are to be put in, the line is temporarily held by means of clamps and turnbuckles, as is shown in Fig. 11. until after it is pulled up and the curves are in position. Then the switch is put in in very much the same manner as the splicing ears and circuit breakers,

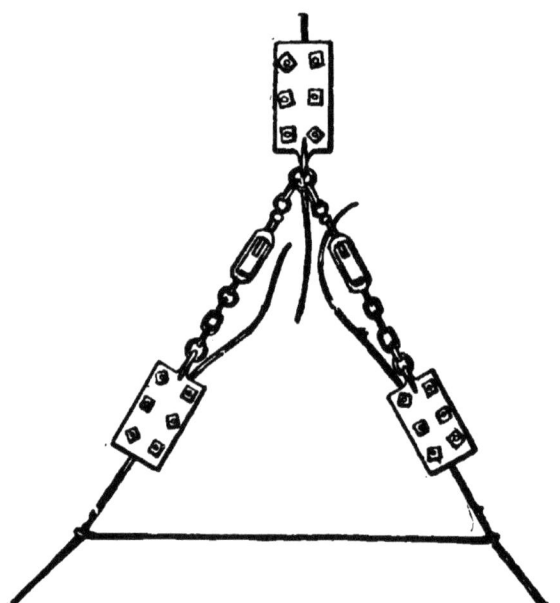

FIG. 11.—TEMPORARY MANNER OF HOLDING CURVE.

enough of the end of the wire being left to allow the switch to be moved should the occasion require.

"Before the insulators are placed in position they should be painted with either P. & B. paint or with Bonnell's enamel, which will lessen to a great extent the leakage due to moisture. These insulators should also be of sufficient size to obviate any possibility of an arc being formed by the trolley running off and striking both line wire and hanger. The permanent pull-off for the curves should be made of cable wire, and should not be fastened to the ear bodies permanently until the road has been in operation for some time, as they are liable to stretch some one or two inches.

"If the trolley runs off at any switch it should be chalked and the car run under it a few times, when it will be seen just where it runs off, and the switch moved forward or backward will remedy the trouble if the switch is properly constructed.

"At every tenth pole short pieces of No. 6 B. & S. wire should be run from the feed wires to the trolleys and there fastened by regular feed ears, and the trolleys should also be cross connected at short intervals.

"Broken or otherwise defective insulators may be found where iron poles are used by a wire leading from the base of the pole to the rail, a flash due to the resistance of the current indicating a ground. Slight ones give a very small shock, but a man standing on the rail holding the wire will receive a very decided shock.

"A mixture of black lead and vaseline rubbed on the trolley wire will prevent the clinging of sleet, and will not clog the trolleys unless too thickly applied. Breaks in the line may be temporarily repaired by means of a short section of wire clamped to the trolley and pulled up with a turn-buckle; after which the permanent repairs may be made by splicing in a section of wire with regular splicing ears."

TRACK.

In an article by Mr. C. J. Field, he states: "The track of street railway companies before the introduction of electricity was more behind the times than any other part of their equipment. The old flat rail is antiquated and antedated, and in a few years its use will be obsolete. The necessities of electric railway traction—in fact, of any traction—have impressed upon the street railway companies in their equipments the requirement of a good roadbed for the successful operation of a road, and we find this part of the problem receiving as much attention as any with companies who appreciate fully the work before them. The general construction to-day is girder rails of from 60 to 80 pounds per yard, placed on chairs where block paving is in use, with ties 2½ to 3 feet between centres. We find in some cases even 90 and 100 pound rails used, but we believe in more moderate weight for the rails and the ties placed closer on centres. We believe this has been the general experience in railway work. Such a style of construction costs from $9,000 to $10,000 per mile. In suburban roads, on streets where there is no paving, we find the T rail being used; the roadbed can be properly constructed on this basis with 45 to 50 pound rail, for $6,000 to $6,500 per mile, the rail being spiked directly to the ties."

Mr. Beckley states: "Those who propose to substitute electric for horse power will make a great blunder if they attempt to put in cheap construction or material. We who have gone into this matter have learned that the track upon which it is proposed to operate electric cars should be of girder or T-rail, of not less weight than 50 pounds to the yard of T, and 62 pounds to the yard of girder rail. The weakest place in the track is, of course, at the joint, and no cheap contrivance at that point should on any account be permitted. With girder or T-rail construction it is, it seems to me, a useless expense to lay a continuous supplementary wire. The rails should, of course, be well and heavily bonded at the joints with iron, not copper, wire, and cross connection of rails be frequently made. Where tram rail track is used I think a continuous wire should be laid and connected with the bond wires."

In the article by Mr. Colgate before referred to the following instructions were given regarding the bonding of the track:

"In the successful operation of all single trolley electric railroads it is necessary to provide suitable means for maintaining a complete metallic circuit, which is generally accomplished by connecting one pole of the generator (in most cases the positive one) to the rail or supplementary return wire, the other pole being connected to the trolley wire.

"Many attempts have been made to operate roads on what is known as a grounded circuit; that is to say, by using the earth for the return circuit. This practice, although advocated by some—one man even going so far as to apply for patents on the details of this form of construction—is, in the writer's opinion, poor policy, except as a makeshift. In case of necessity, it may be done by driving bars into the soil in the vicinity of the tracks, to which the rails are connected by wires, or by connecting the rails to the remains of some defunct tin roof, which should be buried in the moist earth.

"The wire leading from the dynamo should be connected to a similar ground plate, which ought to be about 20 feet square. This was done in the case of the Essex Electric Road in Peabody, Mass., and has proved quite successful. I understand that this method of grounding is to be done away with on this road and bonding substituted at an early date.

"One of the first methods of connecting the rails was to drive copper wedges between their ends, but this did not give satisfaction, owing to the expansion and contraction of the rails. The method which came into use was that of connecting each rail to the rail next it by means of a copper or iron bond, which should be of sufficient sectional area to carry the current with almost no loss, No. 6 copper wire being the size in general use. At every fifth rail the two sides of the track should be united by means of a No. 0 copper wire, and in the case of a double-track road both tracks should be connected at about every tenth rail by a similar bond. The object of cross-connecting the tracks and rails is to insure a continuous circuit should any of the bonds become broken by the settling of the pavement or of the rails.

"In case of the track being of the ordinary stringer type, and already laid, it is best drilled by means of an upright drill clamped to the rail and operated by hand. The drill I have in mind is very strong and light, and when operated by two ordinary laborers it will bore a 3-inch hole in ⅝-inch iron in about four minutes, including the time to set it in position. The best mode of operating is to raise the end of the rail and insert under it a block of wood about 12× 3×3 inches to keep it in place. On no account should the men be allowed to use stones for this purpose, as they are liable to slip out at the most inopportune moments, and the spring of the rails is quite sufficient to amputate a finger or two with surprising quickness. After the end of the rail is secured in position the drill is fastened to the rail by means of a clamp and set screw. I may say here that a spanner should always be used in connection with this rail, as a monkey wrench soon becomes so clogged with sand as to render it useless. A hole is then drilled through the thinnest part of the rail about three inches from the fishplate.

"The driller then proceeds to drill all the rails in rotation. At every fifth joint he drills a hole two feet from the fishplate for the reception of the long bonds. A corresponding hole is also drilled in the opposite rail. At every tenth rail, if it be a double track road, an extra hole is drilled in the inside rail for the other cross bond. I consider it the wiser plan to have enough drillers to enable them to keep a distance in advance of the rest of the gang, for various reasons, the most prominent one being that the drills suffer less from being clogged with dirt.

"After being drilled the holes should be counter-sunk to the depth of about one-eighth of an inch by means of a countersink and carpenter's brace. One of the best tools for this purpose is made like that shown in Fig. 12. Such countersinks can be made by any blacksmith, and are easily sharpened. It is necessary to keep them very sharp, as they are absolutely useless if dull.

"After the holes are countersunk the paving between rails at the joints is removed for the space of about two feet, and an excavation made just wide enough to enable a man to dig out the soil to a sufficient depth to expose the ties. The men engaged in this work labor to the best advantage in sets of three, two loosening the paving, while the

FIG. 12.

third removes it. At the extra holes for the cross bonds a small trench should be dug, uniting them.

"On the completion of the digging the bonds are riveted in place. This should be done by men working in pairs, one man raising the rail, while the other does the riveting. The end of the rail is raised by means of a bar, and a block of iron is placed under it parallel with the stringer. This block should be about 8×1½×3 inches. The end of the rivet is then riveted in the hole, and is forced home by allowing the rail to spring down on it. Sometimes it is necessary to pound on the rail with the head of the bar, which will help very materially in forcing the rivet into place. The riveter must exercise care and see that his helper does not cut the wire off between the web of the rail and the riveting plate. While the plate is still in this position the projecting end of the rivet is cut off flush with the rail, and a head formed which fills up the countersinking. The end of the rail is then raised again, the block removed and the stringer cut away sufficiently to allow one-half inch clearance between it and the head of the rivet, which gives the rail a chance for free movement without danger of severing the bond between the stringer and the web of the rail.

"The rivet in the other end of the bond is fastened to the adjoining rail in the same manner. After this the bond is bent down alongside the stringer, held in place by a tinned iron staple, and the paving replaced. The cross bonds should be buried deep enough to be out of danger of being broken by the settling of the pavement. All bonds should be of sufficient length to allow a slight movement of the rail caused by expansion, loose spikes, the passing of cars and other vehicles. The writer prefers those bonds which are made of one piece of copper, as they are not so liable to cause trouble by the rivet becoming loosened from the side. There is another excellent bond made by having a hole in the rivet near the head just large enough to contain the wire, which is then upset in such a manner as to cause it to jam, making it impossible for it to pull out. (See Fig. 13). If the bonds are made by twisting the wire around the rivet under the head and then soldering it, it is good policy to examine them and see that the rivets do not taper near the head under the wire; also make sure that the wire is up close to the head of the rivet. If they do not comply with these requirements they should be rejected as worthless, for they will cause endless trouble by the wire breaking loose. The rivets should be about one-fourth of an inch longer than the thickness of the rail for which they are intended, the projecting end being cut off as has been described, and should be

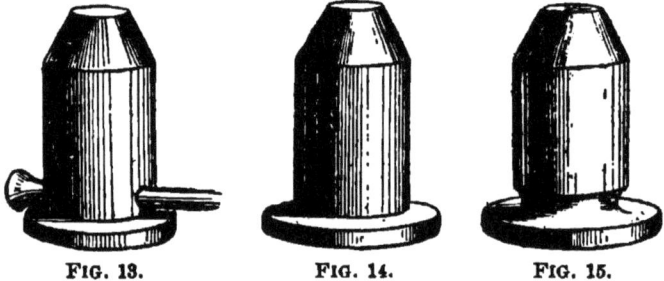

FIG. 13.   FIG. 14.   FIG. 15.

one thirty-second of an inch larger than the holes in the rail.

"Figs. 14 and 15 show about the correct shape of the rivet. Attaching bond wires to the track by a hole drilled partly through the rail, into which the wire is inserted and held in place by a dowel pin, has rather fallen into disrepute owing to the wire breaking off close to the rail when the paving is replaced. But there is no reason why this arrangement should not be used on suburban roads where there is no paving and the road is used not for a highway, though care should be taken not to cut the wire when inserting the pin.

"All methods of bonding without the use of a supplementary return wire are generally avoided for various reasons, the most prominent one being that the breaking of a number of bonds caused by the removal of rails will produce trouble. But in this case a short length of wire might be run temporarily to take the place of the rails which have been taken up during repairs.

"Frogs, switches, and other castings should be connected in the same manner as other rails, but should be wired by a small gang of men after the rails are all done.

"These castings should be drilled from the top down, though much difficulty is sometimes experienced owing to the presence of sand and slag. The latter usually is found low down, and is so hard that it will turn the edge of the hardest drill. The best way to drill these difficult castings is to use a Morse twist drill which has been tempered glass hard, which should be turned and fed very slowly, being kept moist with turpentine, to which a little spirits of camphor has been added. After drilling all but one-fourth of an inch take out the drill and put in a punch, which may be struck with a sledge. This treatment will cause the lower coating of slag to give way. Care must be taken not to strike hard enough to fracture the casting. In crossing steam and other railroad tracks, their rails at the point of crossing should also be connected with the track, or, better, with the return wire.

"The most approved manner of bonding track is by means of a return supplementary wire, which should be one No. 0, or two of No. 3, tinned copper.

After the rails are drilled and countersunk in the manner just described, the paving should be removed at the joints in the left hand side of the track between the rails. The openings are deepened enough to expose the ties, or very nearly to that depth. A ditch is then dug next to the right hand stringer on the inside of the track, to a sufficient depth to enable the supplementary wire to be easily stapled to the ties. The bonds are now riveted in place, every alternate bond on the left hand side being long enough to reach across the intervening space to the ditch opposite, where it is afterward connected to the supplementary wire; adjoining left hand bonds are connected, and the joint soldered. The right hand bonds should be about six inches longer than the height of the stringer, which will allow two wraps around the supplementary wire, two bonds being fastened in each rail. After the bonds are riveted in place the supplementary wire is run out along the ditch. This wire should be No. 0 tinned copper (and should be purchased in coils not larger than about 150 pounds each), to which the bonds are connected by wrapping the ends around the wire. Gas tongs are far better than pliers for this work, on account of the latter getting clogged with sand. The joints are now all soldered by pouring hot solder into them, holding a ladle underneath to catch the surplus. The solder must be very hot for this work. A large charcoal furnace is about the best means of keeping it at the proper temperature. All gasoline apparatus the writer has ever seen proved a failure on account of getting clogged with dirt. After the joints are soldered the wire is fastened down in place by means of tinned staples, one being driven into every third tie.

"In bonding track while being laid the holes are drilled and the rivets fastened in before the rails are finally in place. A small hand punch may be advantageously used for this purpose.

"After the rails are placed in position and the ties tamped two No. 3 wires are run out to the stringers and the ends of the bonds fastened to them. These two wires are connected every 200 feet by means of

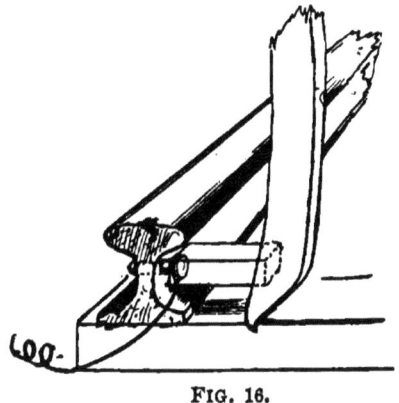

FIG. 16.

short pieces of wire of the same size. The object in using two wires in this case is a saving of copper in the bonds, which is quite an item in a long road.

"Girder and T rails should be drilled through the side, but in the case of rails already laid this is not practicable except in the latter case, where a ratchet drill may be used. The rivets are held in place while being headed by means of a crowbar and block of iron about 6×6×3 inches, and is used as shown in Fig. 16.

"In bonding Johnstown rails already in use, the

drill is clamped to the rail as is shown in Fig. 17, the soil having previously been dug away to a sufficient depth to allow the drill to be easily set in position. To rivet bonds in this type of rail it is necessary to use a bar with a flat turned up end, similar to a claw bar used for the purpose of drawing spikes. This is used in connection with the riveting block

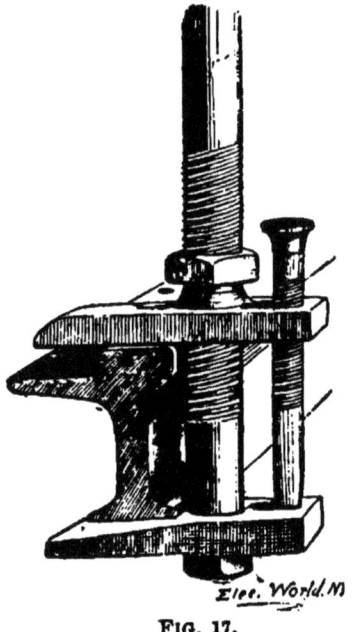

FIG. 17.

as shown in Fig. 18. One man holds the rivet in place while the head is being formed by the other.

"The rivets of broken bonds should be removed with a punch to which a wooden handle is attached. Extra long bonds may be made by twisting the wire around the rivet close up to the head and there securing it by soldering. Tinned rivets are the best for this purpose, but if it is impossible to procure them already tinned this may be done by heating them red hot, plunging them into a saturated solution of sal ammoniac to which a little muriatic acid has been added, and then dipping them while still wet into melted solder. The workman should take great care during this process, as the solder is almost sure to spatter.

"Castings through which it is impossible to drill may be connected to the rest of the circuit by plac-

Elec. World. N Y.

FIG. 18.

ing a copper plate under them which is connected to the supplementary wire, and spiking them down with extra long spikes, or by bolts, to insure good contact. Girder rails are sometimes grounded in a similar manner by means of a sheet of copper between the rail and chair."

TRACTION POWER.

There appears to be no small difference of opinion regarding the co-efficient of traction, that is, the number of pounds pull per ton which is necessary to move a car on a level. These differences are due not so much to inaccurate measurements, but probably chiefly to the fact that slight differences in the track, roadbed, imperceptible grades, and the amount of dirt, affect the results quite materially. Mr. Crosby found by means of a dynamometer that it required about 25 pounds per ton for a car weighing about 12,000 pounds on an average street car track with an ordinary flat rail. Mr. Mailloux found that the traction per ton in some cases reached over 30 pounds and as high as 40. He believes that the traction per ton increases with the weight of the car, stating that it might be 20 to 25 pounds per ton for 3-ton cars, while for 7 or 8 ton cars it might probably be 30 or 35 pounds per ton. This view might explain the lower figure found for the traction of horse cars, as they are of so much smaller weight. Mr. Prescott reports a test made with a dynamometer on a car pulled by men, in which he found the co-efficient to be in the neighborhood of 30 pounds to the ton for a 6-ton car. Some tests made in Paris on the Decauville narrow gauge road were reported to give as low as 5½ pounds per ton, which is extremely low. A test made by Tresca on an electric line between Paris and Versailles gave 27 pounds. A set of numerous careful experiments made by Mr. Hubel on horse car lines in Hamburg gave 33 pounds. Mr. C. J. Field states: "A fair basis for the power, on general conditions, for 16 to 18 foot car bodies, is 20 to 25 horse-power per car, which, with a properly designed and constructed plant, will give the desired power. The cost of generating this power for railway work for 16 and 18 foot cars is three to five cents per mile for all expenses of the generating station. In some roads we find that cars of a larger size than these do not necessarily take a proportionately larger amount of power. We find from practical experience that a car 32 or 33 feet long, double the size of the 16-foot car, takes, under general conditions, about 50 per cent. more power, and we find by the same experience that a trail car adds about 50 per cent. to the amount of work to be done on the motor car for the same size. As to the minimum and maximum amount of power taken on an electric car, we find that a general average for a 16-foot car, under ordinary commercial conditions, without excessive grades, is one horse-power per car mile per hour; or, a car operating at an average 10 miles per hour means an average of 10 horse-power per car. This same car will give, however, on a load diagram, taking all its conditions, from maximum to minimum, a variation of from nothing to 50, 60, or even 80 horse-power. This gives us an idea of the severe strains and conditions to which an electric motor is subjected. Under general conditions, 30 horse-power, with two 15 horse-power motors, has been found about right; in fact, we even find the companies tending toward a larger installation of power, particularly when using larger than a 16-foot car body, and we find to-day, being installed for rapid transit in inter-suburban work, 40 and 50 horse-power electric equipments per car, many of them operating at a speed of 30, and even 40 miles per hour. As the amount of power is directly proportionate to the speed, we can readily see the requirements for such an amount of power."

Regarding the starting effect, Mr. Crosby reports a test in which he found it took 14 ampères to start an ordinary car with Sprague motors, on a level, rusty track; as the current was cut down by a resistance, this does not mean that the voltage was the full 500. He checked it with a dynamometer and a rope and found it took 175 to 200 pounds for 5 tons, or about 35 to 40 pounds per ton, which he said checked exactly with the calculated torque resulting from the 14 ampères. Mr. Mailloux found that the initial traction was as high as 80 to 100 pounds per ton.

SPEEDS, GRADES, LOADS, ETC.

Regarding the speed of electric cars, Mr. F. L. Pope states: "The average speed of the horse car is about six miles per hour. The question is sometimes asked, how fast may electric cars be safely run in a city street? One fact within my own knowledge will go far to answer this question. There is a heavily traveled street in Pittsburgh only 36 feet wide, containing a double-track cable road, which leaves not more than nine feet space on each ' side. At first the cable cars were run at the rate of seven miles per hour; afterward the speed was increased to $9\frac{1}{2}$ miles per hour. The records show that there are not so many accidents under the present arrangement as there were before. Pedestrians and drivers are more careful and take fewer chances. The schedule rate of the electric cars in Cleveland is nine miles per hour, and in some parts of Boston as high as 12. The value of an electric railway to the public is largely determined by its speed, but the economical aspect of the question is equally important. If we make six miles per hour with horses, and nine with electricity, each car does 50 per cent. more work without increased expense for conductors' and drivers' wages, which is an important item. Another economical feature due to the use of electricity is the ability to haul one, or even two, tow cars without loss of schedule time on special occasions, when the traffic is unusually great. Nothing is more astonishing than the capacity of the electric cars to make their schedule time in the face of the heavy storms of a New England winter. It is a common sight to see an electric car running, apparently with perfect ease, up a heavy grade through snow a foot deep, pushing or pulling other cars loaded to their utmost capacity."

From the reports of a number of companies compiled by Mr. Mansfield, he finds the following: "The average speed of all the roads is 8.7 miles per hour. The maximum is 30. The average grade is 6.7 per cent., and but 12 roads report as having none, or very small ones. The maximum grade is 13½ per cent., and this extends for 1,500 feet. The road suffering from such an infliction is in Amsterdam, N. Y. Thirteen roads report 10 per cent. or over. Nashville, Tenn, reports an 11 per .cent. grade for 1,300 feet, and Burlington, Ia., an 8½ per cent. for 1,500 feet, while Wilmington, Del., reports a 7½ per cent. for 3,000 feet. The loads carried up these grades by two 15 horse-power motors are, to say the least, surprising. Amsterdam reports one motor car and 52 passengers. Nashville reports one motor car and 77 grown passengers; Burlington, one motor car and 75 passengers, and Wilmington, Del., reports one motor car towing a disabled motor car. Several roads report as towing one car with both full of passengers up eight and even nine per cent. grades, but for short distances. Auburn, N. Y., reports as having towed five cars, all loaded, with one motor car. The grades in this instance were slight. In all these instances unquestionably the motors were exerting power considerably beyond their rated capacity. Trains carrying 350 passengers have been moved by two 15 horse-power motors; 200 passengers is an every-day occurrence. Surely this is approaching steam railroad practice. Such information is certainly useful to the electric manufacturing companies."

A number of interesting and instructive tests were made by Mr. Charles Hewitt of the speed of electric cars in city streets, from which important conclusions can be drawn. The report is given here practically in full, as published: "It has been asserted that it is impossible for any car operated on the surface of a city street to make real rapid transit, on account of the crowded condition of the streets. A close study of our large cities will show that the principal lines of travel are crowded for only short distances, and that beyond these crowded portions a high rate of speed can be maintained with safety.

"It only remains to be shown, then, what the electric cars can do with safety on the unobstructed portions of their routes.

"Much has been said and written about the speeds of electric cars, but so far as the writer is aware, no speed record of an electric car in actual service has ever been published heretofore. Last summer the writer had the pleasure of making some tests for the Edison General Electric Company, on Niagara street, Buffalo, covering this question, and the result of those tests is shown in the accompanying diagrams. The route extends from Main street, in the heart of the city, to Hertle avenue, by way of Niagara street, a distance of 4½ miles; or 9 miles for the round trip. The schedule time, including terminal stops, is 64 minutes, or an average running speed of 9 miles per hour. Referring to the diagrams, we will see what speed the car actually made in order to fill the schedule requirement. A few words of explanation will perhaps make the diagrams clearer. The record was made by a Boyer railway speed recorder,* belted direct to one axle by means of a flexible wire belt. Every movement of the car, either forward or backward, is therefore recorded. The horizontal lines represent speed in miles per hour as shown at the left hand end of each diagram. The vertical lines represent quarter mile distances.

"The car in each instance was started from the car barn, which is in the middle of the line, and run to Hertle avenue, the suburban terminus, thence back to Main street, the city terminus. Ampère and voltmeter readings were taken every 15 seconds, and when the car was carrying passengers the number on the car was noted every time any one got on or off. The car was 35 feet long over all, was mounted

---

* For a description of this instrument see Miscellaneous Appliances and Accessories.

upon two trucks, and weighed, with the No. 10 and No. 14 motors, about 20,000 pounds, and with No. 12 motors about 22,000 pounds.

"The first test was made with the No. 10 motors; the car being run from the car barn to Hertle avenue terminus and back to car barn without passengers, in order to determine what was the highest speed that could be obtained. As the car had to be run in between two cars in regular service, the result is somewhat less than what it would have been if the track had been entirely unobstructed, but it will be noted that a speed of 21 miles per hour was attained within a distance of 2,100 feet from the full stop. The gradual acceleration of speed shown by the gentle curvature of the line is also of some interest. The remaining runs before reaching the car barn were interfered with by overtaking the car ahead, although a speed of 17 miles was attained. At the car barn the car was put in regular service carrying passengers, and the record speaks for itself. The average speed shown is somewhat greater than the schedule speed, as the time of stops is not recorded. It is interesting to note that in order to make a schedule speed of nine miles per hour it is necessary to make a maximum speed of about 13½ miles, or 50 per cent. above the schedule speed. This record also shows that a car can attain a speed of 15 miles per hour within a distance of 350 feet from full stop.

"The next test was made with the No. 14 motors. These motors were mounted on the same trucks that had been used in the previous test with the No. 10 motors. The car was first run to Hertle avenue

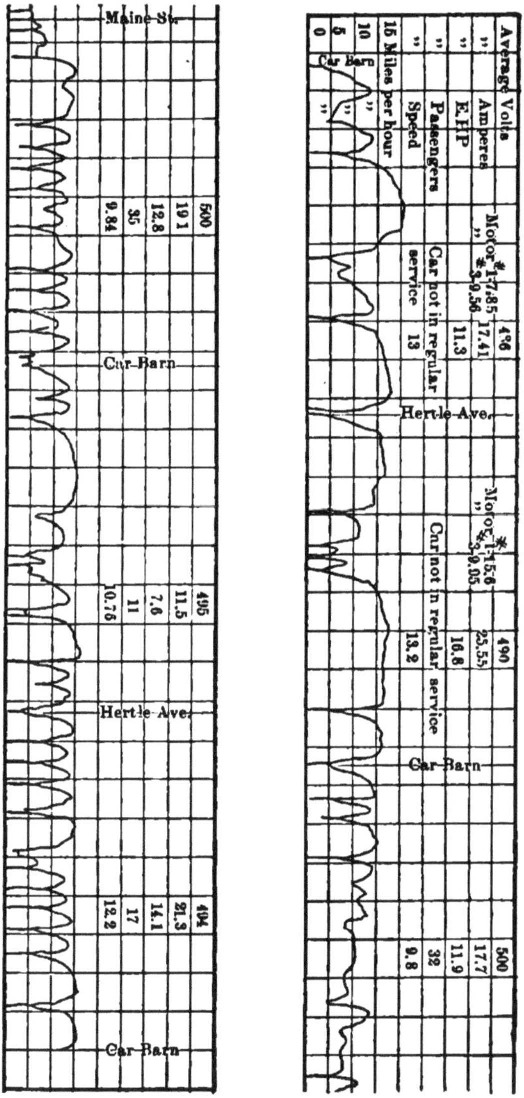

Average Volts		496	490	500	495	494	500
" Amperes	Motor *17.85, **3.56 17.41	17.7	Motor *15.6, **3.05 25.55		11.5	21.3	
" E.H.P.	11.3		16.8	11.9	7.6	14.1	
" Passengers	Car not in regular service 13		Car not in regular service 13.2	32	11	17	
" Speed				9.8			
16 Miles per hour							
10 "						10.76	12.2
5 "		191					
0 Car Barn "	500	12.8					
"		35					
"	9.84						

Maine St. · Car-Barn · Hertle Ave. · Hertle Ave. · Car-Barn

FIG. 19.—TWO NO. 10 MOTORS. AVERAGE LOAD ABOUT 27,000 POUNDS. TRACK NEARLY LEVEL. Average of Total Readings: 494 volts, 18.8 amperes, 12.4 e. h. p., 24 passengers. Average speed without passengers, 13.1; with passengers, 10.4 miles per hour.

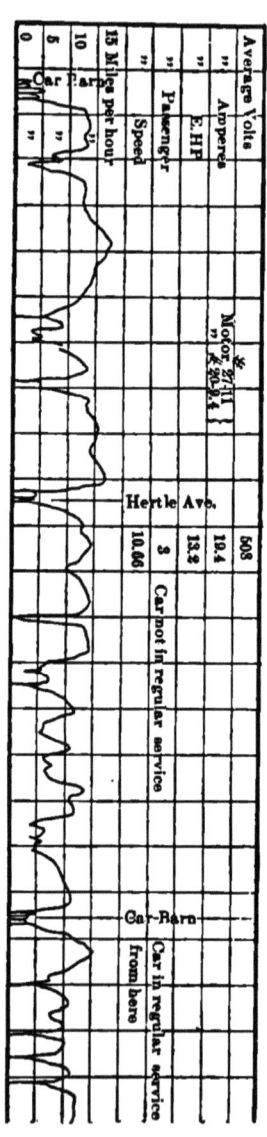

FIG. 20—(See Continuation on next page.)

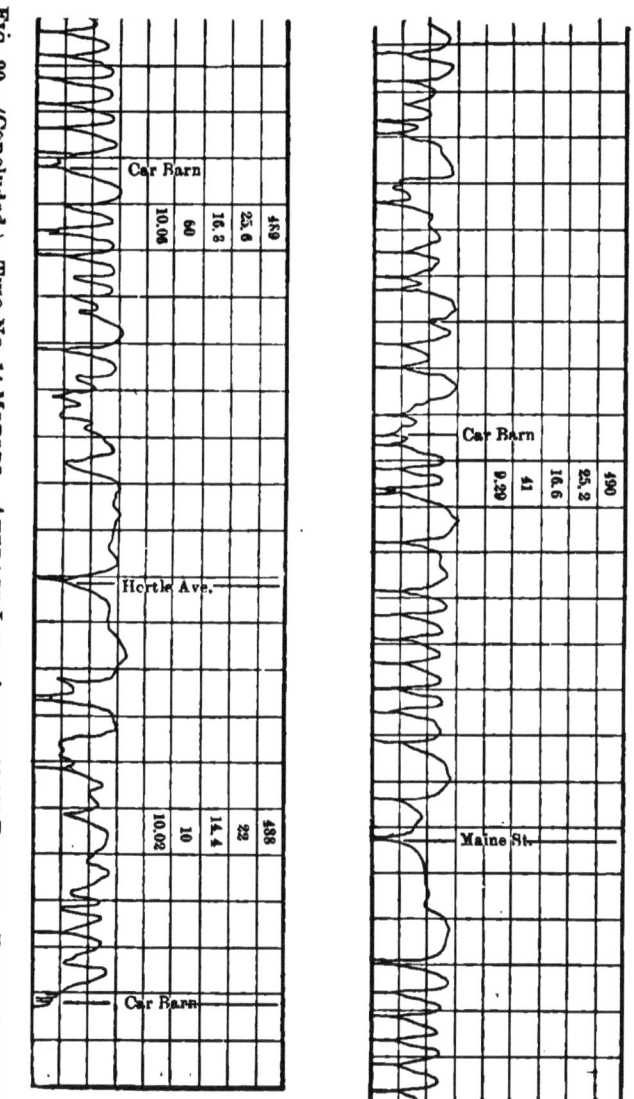

Fig. 20.—(Concluded.)—Two No. 14 Motors. Average Load about 27,000 Pounds. Track Nearly Level.

Average of Total Readings: 494 volts, 22.4 amperes, 14.9 e. h. p., and 35 passengers. Average speed without passengers, 10.63 ; with passengers, 9.73 miles per hour.

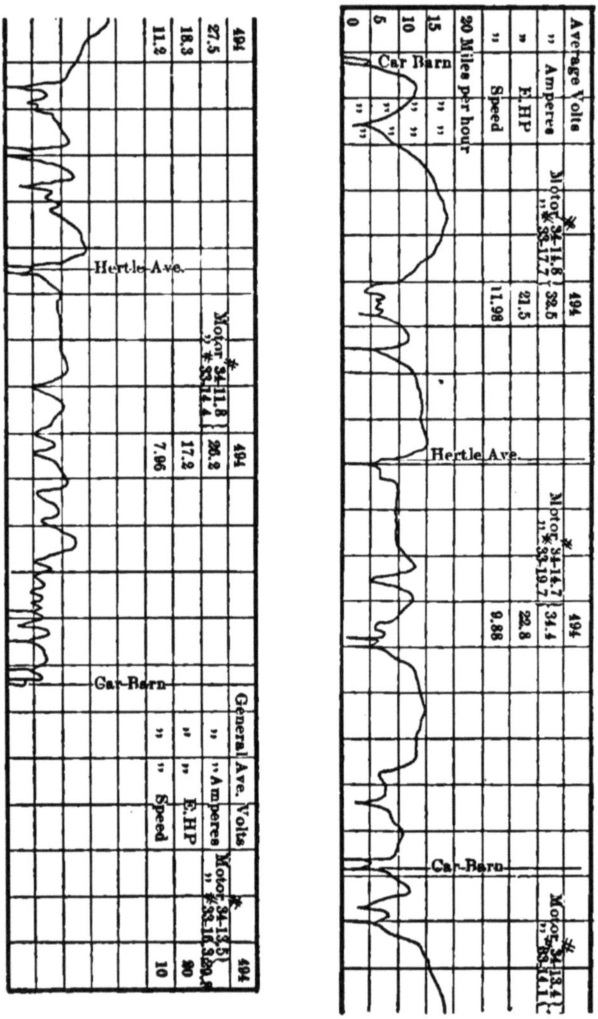

FIG. 21.—TWO NO. 12 MOTORS. LOAD ABOUT 22,000 POUNDS. TRACK NEARLY LEVEL.

and back without passengers, and was then put in regular service. In connection with this diagram I wish to refer to what has been said in the beginning of this article. Although no part of Niagara street is very crowded, still the interference is greatest near the Main street terminus. It will be noted that in every instance between Main street and the car barn the stops are very numerous, for long stretches being one to every block (about 600 feet). Between the car barn and Hertle avenue, however, the stops are comparatively few, and it is here we would expect to get a greater speed, as is distinctly shown by the diagram. It is not so prominent, however, as it would have been had the required schedule speed been 10 or 12 miles per hour, as on some of the long lines in Boston and other cities. It is evident from the first part of this diagram that a speed of $17\frac{1}{2}$ miles could have been maintained for long stretches if necessary, thereby increasing the average speed.

"The third test was made with the No. 12 motors. These motors are larger and more powerful than the No. 10 and No. 14. The same car body was used as in the former tests, but the motors were mounted on new trucks, used for the first time in this test. For this reason the car was not put in regular service. Two trips were made from the car barn to the Hertle avenue terminus without passengers with results as shown in the diagram. Since this test was made this car has attained a speed of 28 miles per hour, and has shown an average ampère consumption of less than 20. With regard to the average ampère readings in all these diagrams, it may be said that they are somewhat higher than they

would be in ordinary practice, due to the long runs
at high speed without passengers, and also to the
fact that the starting current is included in many
readings."

CARS.

The chief matter of interest during the past year,
regarding cars, has been the question of whether it
is more satisfactory to use long or short cars.   The
balance of the opinion appears to be in favor of long
cars. Mr. C. J. Field states: "One of the questions
on which we find more variety of opinion than any
other is what is the best size, type and style of car
for given cases and conditions. The old standard
16-foot car body we find is now being widely
departed from, and the problem is, how large a car
can we get on a single truck with four wheels without excessive destructive effect on the roadbed?
and what is the longest car we can operate on
street car service economically on an eight-wheel
base? We believe the limit is reached with a single
truck in a 20-foot car body ; we know that the truck
manufacturers claim in some cases to operate a
longer body, but we do not believe it wise. An 18
or 20 foot car, running under close headway, we
believe to fulfill best the conditions of city traffic in
the larger cities. Such a car, with a wheel base of
seven feet, and in some cases seven feet six inches,
where curves are not too sharp, will give satisfaction, and not be too severe on the roadbed where
the same is properly constructed. Some companies
have favored the use of a vestibule on street cars.
We believe, though, that any vestibule is a failure

and a misnomer. It accomplishes no good, and causes much trouble. A shield over the dashboard for the motor man in winter would give all that would be required. What is wanted on a street car is that which will allow the freest ingress and egress from the car for the passengers, and anything that retards this—and a vestibule most certainly does— is a detriment and an obstacle to rapid transit. On some roads we have tried the introduction of even larger cars, say, 28-foot body, or 36 feet over all. Such a car, of course, has to be put on a double truck. These cars have found favor with some companies when first considering the problem. The difficulty with them is in getting the passengers in and out of the cars as quickly as possible, and making too many stops, due to the larger number of passengers carried. For inter-suburban heavy traffic, with few stops, we believe such a car would fulfill the requirements, but only in such a case." The same writer states that a 32 to 33 foot car takes, under general conditions, about 50 per cent. more power than a 16-foot car.

The West End Company of Boston find that there is a decided gain in the ratio of the operating expenses per passenger to the earnings in the long cars as compared with the short cars. The weight of a long car empty is about 18,000 pounds, and they find that on the level it takes an almost imperceptibly small amount of additional power to draw this car when loaded with 15,000 pounds besides the weight of the car; on grades, of course, the difference is felt. From this they conclude that the cost of the power for the long cars is very little more than for the short ones, although they carry nearly double the number of passengers; the expense for conductors and drivers, of course, remains the same, while the accommodation of the passengers is considerably greater. They contemplate the general introduction of long cars; 175 are to be added before January, 1892, making a total of 400 long cars in use during the winter of 1891-92. They also contemplate getting double deck cars for the same eason. The long cars earned 44 cents per mile in May and 47 cents in June. The short cars are 16 feet long and are said to cost $4,000 each. The long cars which they have adopted as their standard are 26 to 28 feet in the body and 35 feet over all.

Mr. Beckley, however, differs from the above opinions, as he considers it a mistake to equip a car body of greater length than 18 feet, and thinks a 16-foot car is better still. During the hours of the day when travel is heavy he says it is easy to pull a trailer, and when traffic is light you are not then using up your power in hauling around a "great lumbering double truck structure practically empty."

MISCELLANEOUS.

In a compilation from reports made of a number of companies, Mr. Mansfield finds that 44 roads report as never having been stopped by any cause, 23 were forced to stop because of the steam plant, failure of water, floods or fire, and 26 from electrical troubles, the main cause of these troubles being lightning. "I consider this a very fair showing, and feel confident that as the art advances these tie-ups will grow less and less, and finally become of rare occurrence.

"Out of the total number of 137 roads heard from, only 32 report as having made any tests of either engines, dynamo, or motors, and 53 upon the overhead work. Surely this is lamentable. There is nothing more essential to an electric railroad than a first-class voltmeter, ammeter, galvanometer, and, if possible, a wattmeter. Electric light, telegraph and telephone, and all other electric companies are supplied with necessary testing instruments, and in most instances a most rigid system is maintained. Every railroad should be continually testing its circuits, station and cars for leaks or grounds. By this means, and this means only, can they avoid trouble and consequent damage. Furthermore, for the sake of economy, these instruments should be used freely. Particularly is a wattmeter useful in a power station. I advise, urge, and beseech every company to supply itself with these instruments, and to put them in the hands of a competent person, or if they can afford it, a thorough electrician."

In another similar compilation it was found that the average car mileage of a car was about 101 miles. The maximum reported was on a small road, in which the cars made 175 miles a day. Two other small roads report 160, while a considerable number gave 125 to 150. The lowest was 46.

Mr. Everett advocates the use of an oil headlight that can be removed easily, so that when anything happens, say underneath the car, the oil headlight can be used to better advantage than the electric

light.   For the same reason, he says that one porta-
ble oil light should be kept in every car.   He also
advocates electric brakes, electric fare registers,
and electric heaters, which, he states, are now used
on quite a number of roads.

In a committee report read by Mr. Crosby on
"Standards in Electric Railway Practice," published
in *The Electrical World*, page 328, October 31, he
advocated the adoption of uniformity in rating
motors, defining dimensions of car, in the nomen-
clature of terms and phrases used in electric railway
work, and in the method of keeping accounts.   The
report is accompanied by suggestions for the basis
of such standards, including a list of terms, with
their definitions, and a proposed standard method
of keeping accounts.   The report is well prepared
and should be studied by all those interested in railway work. Want of space prevents us from reprinting it here.

_____

CHAPTER IV.

COST OF CONSTRUCTION AND OPERATION.

Among the matter published was an article by
Mr. J. S. Badger, which we reproduce here in full,
as it contains very interesting and doubtless very
reliable matter, and is probably the best article published during the year on this subject.

After pointing out how difficult it was to secure
sufficient data to be of value for purposes of comparison, Mr. Badger states:

"The data herewith presented concerning electric
roads have been very carefully collected, and will
enable accurate conclusions to be drawn respecting

the relative merits of horses, cable and electricity as a motive power for street cars. The electric roads, which really control the results, have been in operation since the earlier days of the electric railway, and, having passed the experimental stage, their operations are well settled and uniform. In several cases very considerable changes are in progress, but they are such as will conduce to still greater economy of operation.

"The elements which enter into a consideration of the questions at issue are:

"First cost of road and equipment.

"Operating expenses per passenger carried.

"Ratio expenses to receipts.

"Operating expenses per car mile.

"Upon the first and last must rest the decision as to what is the most economical power for street railways.

"The element of first cost may or may not decide the question at once. If the capital available is limited to the amount necessary for the least expensive construction and equipment, this settles the question of choice of motive power. If the capital is limited only by the ability of the road to pay a reasonable return upon the investment, the question becomes more complex; and whether a cheap or expensive construction shall be adopted depends upon whether the interest charge added to cost of operation will be large or small when divided by the total number of units of comparison.

"'Passengers carried per mile of route' or 'per mile run' relates chiefly to density of traffic and consequent value of the franchise, and has only a distant bearing upon the question of operating expenses.

"Cost per passenger carried, and ratio of operating expenses to receipts, considered in connection with fixed charges, relate to the dividend paying ability of the road, or the value of its securities as an investment, bearing in mind the fact that two roads may show the same ratio of operating expenses to receipts, and one be able to pay twice as much as the other in dividends. Therefore these elements do not necessarily have any bearing upon the actual economy of operation, or give any indication as to which of any two or more systems is the most economical.

"Cost per car mile, for cars of about equal carrying capacity, seems to be at present the only basis of comparison. This expense, within the limits of traffic for which the power plant and equipments are adapted, remains pretty constant, regardless of variations in amount of traffic, but this is directly affected by change in value in any item of operating expense just to the extent that such variation is part of the whole expense.

"Whether a road shall show a high or low cost per car mile depends upon its physical characteristics, the motive power employed, to a certain extent upon whether cars are run singly or in trains, and upon the ability of the man responsible for its mechanical operations.

"Whether it shall show a high or a low cost per passenger carried depends to a certain extent upon the foregoing, but principally upon whether few or many can be induced to ride upon it, and therefore upon good management. A careful examination of the roads in question shows the incorrectness of the statement that those 'which have the least expense per car mile have the greatest expense per passenger carried.' There is no uniformity in this respect one way or the other, as between different roads. Theoretically, the expense per car mile would slightly increase with an increase in traffic. In a general way, and without attempting to produce any proof in support of this opinion, it may be said that, other conditions remaining the same, the expense per car mile would increase about as the cube root of the number of passengers carried. In this way, and under conditions seldom realized in practice, the foregoing statement might be true.

"Expense per car mile, for an electric road, depends chiefly upon cost of fuel, efficiency of steam and electric plant, wages, character of road as relates to grades and curves, and last, but by no means least, intelligent care of machinery, etc. Without a knowledge upon these points, a simple statement of expenses per car mile conveys little definite information. With coal varying from $1 to $3.80 per ton, its consumption from 4.3 pounds to 12.2 pounds per car mile, a station output from 3.7 to 8.4 and even 10.7 electric horse-power per car in operation, and wages of conductors and motor men from 10 to 20 cents per hour, the importance of these data is at once evident.

"Regarding care of machinery, a comparison may not be amiss. In the case of one road, an average of one-sixth of the armatures in use go into the repair shop every month, generally because burned out. Another road using the same system, operating more cars upon heavier grades, has not had an armature burned out in nearly a year. As both do their own repairing, the manufacturing company is relieved of responsibility for this difference. Many similar examples could be given.

"In another instance, upon a road of less than 10 miles, there was for months a constant leakage of 20 to 30 ampères on the line.

"While, therefore, the manufacturing companies may have been guilty of sins of omission, the operating companies have been guilty of sins of omission and commission. Experience has been costly for both. Electrical apparatus must have intelligent care, or the repair bills soon assume large proportions, and this frequently causes railway companies, who do not understand the true cause of the trouble, to condemn electricity as an expensive motive power. What can be and is actually accomplished in practice is shown elsewhere in this paper. In isolated cases circumstances may strongly favor some one of the three systems under consideration to the exclusion of the others. This must be determined by expert examination; but whichever shows the best results out of a large number of cases must stand at the head, as the most desirable and economical motive power for street railways.

"The item of first cost is the subject of considerable discussion. Direct information concerning cable roads has not been obtainable; but as the figures we cite are those given by the Census Department, and do not seem to have been questioned by any authority upon the subject, they may be accepted as substantially correct. Our data concerning horse roads, being taken from sworn reports to the Massachusetts Board of Railroad Commissioners, can also be relied upon. The figures given concerning investment in electric roads have come from official 'sources, and are confirmed by private information. These are, however, excessive. Most of the roads mentioned were formerly horse roads, and to the original investment has been added the cost of change of motive power; and in almost every case the amount now charged to permanent investment is far in excess of what it would cost to renew the entire power plant, track and equipment.

"Estimates upon the track construction differ greatly, but the limit of profitable investment is not likely to exceed $10,000 per mile; while as fine and substantial a roadbed as electric car ever ran over was built at a cost, exclusive of paving, of about $5,000 per mile.

"The overhead structure need not cost to exceed $2,500 to $3,000 per mile of single track for best wood poles, or $3,500 to $5,000 for iron poles. For double track, iron poles, it would vary from $4,500 to $6,500 per mile, centre pole construction being the cheaper, and in many other respects preferable where it can be adopted.

"From $3,000 to $3,500 per car is a liberal estimate for 16-foot to 20-foot cars, fully equipped, and the average is about two cars per mile of road.

"An allowance of 15 to 20 horse-power per car. at $80 to $100 per horse-power for station equipment,

including steam plant, but not real estate or buildings, is very liberal.

"Thus we have, as an extremely liberal estimate,
$26,000 per mile, exclusive of real estate, buildings
and paving, for a road suitable for the heaviest
metropolitan traffic. And it is a fact that a good
and satisfactory road can be built and equipped for
$20,000 per mile.

"As records of actual experience a careful examination of the statistics presented is invited, and
especially of those relating to electric railways. The
greater part of these has been obtained by personal
visits to the roads reported.

Comparison of Investment and Operating Expenses.

TABLE I.

	Total investment real estate, road and equipment.		Car miles run per annum per mile of street length.	Passengers carried annually per mile of street length.	Passengers carried per car mile run.
	Per mile of street length.	Per mile of track length.			
* 22 Electric roads..........	38,500	27,780	76,158	237,038	3.10
† 45 Horse roads	33,406	31,093	43,345	251,816	5.81
‡ 10 Cable roads.............	350,325	184,275	309,395	1,355,965	4.38

"Table I. shows that, taking street length as the
unit of comparison, in the cases of the roads under
consideration, the total permanent investment of
the electric is only 15 per cent. more than that of

* Car miles run per annum, 14.013,187; passengers carried per annum,
43.614.972; street length, 184 miles; track length, 255 miles.
† All the roads in Massachusetts operated exclusively by horses for
1885-90. Average for six years.
‡ From Census Bulletin No. 55.

the horse roads, while the cable roads cost more than nine times as much as the electric roads. The average speed of cable and of electric cars is about the same, consequently the cable roads ran about four times as many cars per mile of street length as the electric. This would be expected, as the cable roads generally occupy the routes of heaviest travel. The horse roads ran more cars than the electric, for an equal length of road, but the latter, having an advantage in higher speed, greatly exceed in car miles run. The electric roads carried fewest passengers per car mile, but carried as many per mile of street occupied as the horse roads. On account of their more favorable location, the cable roads exceed both the others in passengers per mile of route. The column showing passengers carried per mile run gives a general idea of the relative number of passengers on a car at any one time.

TABLE II.

	Operating expenses per car mile run.	Interest charge per car mile at six per cent. on total investments.	Total of operating expenses and interest per car mile.	Cost per passenger carried, interest excluded.	Cost per passenger carried, interest included.
	(Cents.)	(Cents.)	(Cents).	(Cents.)	(Cents.)
Electric roads	11.02	3.03	14.05	3.55	4.53
Horse roads	24.32	4.62	28.94	4.18	4.98
Cable roads	14.12	6.97	20.91	3.22	4.77

"Table II. shows operating expenses per car mile, all taxes and fixed charges excluded, for each of the three systems; interest charge per mile at six per cent. upon the total permanent investment; total of operating expenses and interest per car mile; cost per passenger carried, interest charge excluded, and the same with interest charge included. Upon every point, save the one unimportant one of cost per passenger carried (interest excluded), the superiority of the electrc road is plainly evident.

TABLE III.

	Ratio of investment, per mile of street length.	Ratio of car miles run annually per mile of street length.	Ratio of cost of operation per car mile, interest included.	Proportional traffic that must be done, per mile of street occupied, to pay operating expenses and 6 per cent. on the investment.
Electric roads	1.152	1.757	.485	.852
Horse roads	1.000	1.000	1.000	1.000
Cable roads	10.486	7.138	.722	5.154

"Table III., for greater convenience in comparison, shows the ratios of the three most important items, and the proportional traffic that must be done, per mile of street occupied, for each system, to pay operating expenses and six per cent. on the investment. Here, more than anywhere else, the superiority of the electric road is plainly evident, the last column showing that in but few cases can there be even a question as to which system offers the greatest inducement to the investor.

SEVEN REPRESENTATIVE ROADS, OPERATED ENTIRELY BY ELECTRICITY.

Road.	Length.		Passengers carried annually per mile of road.	Number of cars in daily operation.	Average daily mileage per car.	Average number of passengers daily per car.	Passengers carried per car mile.	Operating expenses per car mile.	Operating expenses per car per day.	Cost per passenger carried.
	Of all tracks.	Of road.								
1	51.0	35.0	*162,857	50	100	313	3.13	12.29	12.29	3.93
2	40.0	19.5	487,582	140	91	188	2.06	7.80	7.10	3.79
3	16.0	10.0	199,000	16	125	343	2.75	8.43	10.54	3.07
4	8.5	5.0	*460,000	20	83	318	3.82	11.82	9.80	3.09
5	15.5	14.0	167,511	18	106	357	3.35	11.00	11.70	3.28
6	28.0	23.5	286,852	31	108	597	5.51	12.74	13.76	2.31
7	3.8	2.8	200,000	5	92	307	3.33	8.49	7.81	2.55
	162.8	109.8	....	280	......	......	......	9.83	......	3.28

Total annual car mileage, 9,862,000. Total number of passengers carried annually, 29,144,000.

"If the electric roads carried as many passengers per car mile run as the horse or cable roads, which they could easily do, and allowing for the increase in operating expenses due to increased traffic, the cost per passenger carried would be as follows:

Per passenger.
Cents.

At 5.81 passengers per car mile (number carried by horse roads) the cost would be, interest charge excluded.................... 2.38

At 5.81 passengers (as above) cost would be, interest charge included... 3.82

At 4.38 passengers per car mile (number carried by cable roads) the cost would be, interest charge excluded.................... 2.82

At 4.38 passengers (as above) cost would be, interest charge included... 3.51

"This is not in any way an attempt to decry the cable system, as it is undeniable that it has a place of its own, where it is satisfactory to the public and profitable to the investors; but the claim is that, under all circumstances, the electric road can handle just as heavy traffic, as readily and satisfac-

* Estimated.

torily to the public, with much greater economy in operation and much less investment of capital."

Mr. Badger's paper concludes with the following statistics of electricity.

Operating Expenses of Electric Roads.

"Average of 22 trolley roads. Length varying from 3 to 51 miles; cars in daily operation, 3 to 140; daily mileage per car, 80 to 150; average daily mileage per car, 110.

	Expense per car mile. (Cents)		
	Highest.	Lowest.	Av'r.
Maintenance of roadbed and track..................	1.86	.10	.54
Maintenance of line......................................	.95	.01	.12
Maintenance of power plant, including repairs on engines, dynamos, buildings, etc..................	.86	.05	.36
Cost of power, including fuel, wages of engineers, firemen, dynamo tenders, oil, waste, water, and other supplies..	4.95	.48	1.96
Repairs on cars and motors..........................	5.24	.59	1.80
Transportation expenses, including wages of conductors, motor men, starters, and switchmen, removal of snow and ice, accidents to persons and property, etc.......................................	9.47	2.74	4.98
General expenses, including salaries of officers and clerks, office expenses, advertising, printing, legal expenses, insurance, etc.....................	2.95	.79	1.26
Total..	*22.99	*7.80	11.02

Cost of coal varies from $1 per ton for slack, to $3 for r. o. m. (run of mine), and $3.80 for lump.

Wages of conductors and motor men vary from 10 cents to 20 cents per hour.

Consumption of coal varies from 4.3 pounds of slack per car mile to 12.2 pounds r. o. m. per car mile.

The station output varies from 3.7 e. h. p. (electrical horse-power) to 8.4 e. h. p. per car in operation, for roads equipped with 16-foot cars and Edison motors. In the latter case the road had many heavy

---

* Respectively the highest and lowest total for any one road.

grades and sharp curves. One road, equipped with 30-foot double-truck cars (weight complete about 10 tons), six Edison and 14 Short double 15 h. p. equipments, traffic medium and grades moderate, required an average of 10.7 e. h. p. per car in operation.

The best station performance is one e. h. p. for every five pounds of slack or four pounds of nut consumed; and evaporation of 7½ pounds of water for every pound of slack consumed. Return tubular boilers, Murphy furnaces, Armington & Sims high-speed, single cylinder, non-condensing engines, and Edison generators are in use.

Detailed Distribution of Operating Expenses.

For roads of 10 or 15 miles and upward, operating 20 or more cars per day, averaging 105 to 110 miles each, grades moderate, a careful distribution of expenses, based upon the experience of the best roads, will average about as follows:

	Expenses per car mile. (Cents.)
Maintenance of roadbed and track	.54
Maintenance of line	.12
Maintenance of power plant:	
Repairs on engines and boilers	.180
Repairs on dynamos	.101
Miscellaneous repairs	.078— .36
Cost of power:	
Fuel	.868
Wages of engineers and firemen	.653
Wages of dynamo tenders, etc	.232
Oil, waste, water and other supplies	.218—1.96
Maintenance of rolling stock:	
Repairs on motors (extra gearing)	.695
Repairs on gearing and trolleys	.594
Repairs on car bodies and trucks	.512—1.80
Transportation expenses:	
Wages of conductors and motor men	4.262
Wages of starters, switchmen, track sweepers, etc	.268
Cleaning and inspecting cars	.238
Oil, waste and other supplies	.063
Accidents to persons and property	.061
Miscellaneous	.068—4.98

Expenses per car mile.
(Cents.)

General expenses:
Salaries of officers and clerks.. .743
Office expenses.. .138
Advertising and printing... .061
Legal expenses.. .068
Insurance... .161
Miscellaneous.. .091—1.26

Total.. 11.02

Some Representative Roads.

ROAD NUMBER ONE.

BOILERS.—Four H. T.; 16x66, with 54 4-in. tubes.

FUEL.—Nut and slack, at $2 per ton. Consumption, 5.7 pounds per car mile.

ENGINES.—One 250-h.p. Corliss, one 150-h. p. Ball, one 150-h. p. Brown. Generators driven from countershafts, except from Ball engine.

GENERATORS.—Two No. 32, four No. 20. Edison.

ROADBED.—Tram and centre-bearing and side-bearing girder rail. Weight, 45, 52, and 56½ pounds. Four miles paved. Railway company do not pave or do repairing. Length of all tracks, 8.5 miles; street length, 5 miles. Generally good condition.

GRADES.—Steepest, 13.2 per cent.; length, 100 feet. General character of road moderately heavy.

CARS.—Total number in equipment, 36 motor; length 14 and 16 feet. Average number in daily use, 20. No trail cars used.

CAR MILEAGE for one year, 601,966; daily average, 1,663; daily average per car, 83.15.

WAGES.—Conductors and motor men, 15 cents per hour; engineers, $100 and $90 per month.

	Expenses per car mile. (Cents.)
Maintenance of roadbed and track*	1.05
Maintenance of line	.07
Maintenance of power plant:	
Repairs on engines	.035
Miscellaneous repairs	.016— .05
Cost of power:	
Fuel (nut and slack, at $2.00 per ton)	.671
Wages of engineers and firemen	.591
Oil and waste	.063
Water (at 7 cents per 1,000 gallons)	.092
Other supplies	.010—1.43
Maintenance of rolling stock:	
Repairs on motors (ex. gearing)	.917
Repairs on gears and trolleys	.224
Repairs on car bodies and trucks	.501
New attachments	.204—1.85
Transportation expenses:	
Wages of conductors and motor men	4.635
Wages of starters, switchmen, and track sweepers	.318
Removal of snow and ice	.137
Accidents to persons and property	.026
Cleaning and inspecting cars	.298
Oil, waste and other supplies	.032
Animals, registers, and sundries	.268—5.72
General expenses:	
Salaries, insurance, etc	1.532
Office expenses	.041
Advertising and printing	.051
Legal expenses	.004
Miscellaneous	.018—1.65
Total	11.62

ROAD NUMBER TWO.

BOILERS.—Four H. T., 14x78.

FUEL.—Nut and slack, at $1.75 per ton. Consumption, 11 pounds per car mile.

ENGINES.—Two 125-h. p. Taylor-Beck; one 175-h. p. Taylor-Beck; and (put in July, 1891), one 250-h. p. Armington & Sims. Steam pressure, 80 pounds.

GENERATORS.—Five No. 32, Edison, belted direct from engines.

ROADBED.—25, 30, 40 and 56 pound T rail; 35,

* Including the entire reconstruction of one and one-half miles of track with new rail.

DETAILS OF OPERATING EXPENSES.

	Expenses per car mile. (Cents.)
Maintenance of roadbed and track...................................	.95
Maintenance of line..	.20
Maintenance of power plant :	
Repairs of engines, dynamos, etc..................................	.39
Cost of power :	
Fuel (r. o. m., at $2.68 per ton).................................	1.643
Wages of engineers and firemen...................................	.425
Oil, waste, and other supplies...................................	.204—2.27
Maintenance of rolling stock :	
Repairs on motors (ex. gearing)..................................	1.210
Repairs on gearing, car bodies, and trucks.......................	.852
Machine shops and mechanics......................................	.943—3.01
Transportation expenses :	
Wages of conductors and motor men................................	4.473
Accidents to persons and property (insurance)....................	.190
Cleaning and inspecting, oil, waste, etc., included in machine shops and mechanics...	4.66
General expenses :	
Salaries of officers and clerks..................................	.460
Office expenses..	.162
Miscellaneous..	.188— .81
Total...	12.29

ROAD NUMBER FOUR.

BOILERS.—H. T., with Murphy furnace.

FUEL.—Slack, at $1.00 per ton. Consumption, 4.3 pounds per car mile.

ENGINES.—Three 200-h. p. and three 125-h. p. Armington & Sims, belted direct to generators. Steam pressure, 100 pounds.

GENERATORS.—Six No. 32 and six No. 16. Edison.

ROADBED.—Sixty-six pound girder rail. All paved. Railway company pave and maintain 16 feet in width of street. Length of all tracks, 40 miles; street length, 19½ miles. In good condition.

GRADES.—Steepest, 5 per cent. ; length, 400 feet. General character of road, level.

CARS.—Total number in equipment, 85 motor, 140

trail. Length, 16 feet. Average number in use daily, 70 motor, 70 trail.

ANNUAL CAR MILEAGE.—4,625,636 miles; daily average, 12,776; daily average per car, 91.1.

WAGES.—Motor men, $2 for 12 hours. Conductors, $1.75 for 12 hours first year, $1.90 after first year. Shop men $1.50 to $3 for 10 hours.

DETAILS OF OPERATING EXPENSES.

	Expenses per car mile. (Cents.)
Cost of power:	
Fuel	.218
Wages of engineers and firemen	.152
All station repairs and supplies	.111— .48
Maintenance of rolling stock:	
Repairs on motors and gearing	.918
Repairs on car bodies and trucks	.442—1.36
All other operating expenses	6.96
Total	7.8

ROAD NUMBER FIVE.

BOILERS.—H. T., two 78x16.

FUEL.—R. o. m. at $2.15 per ton. Consumption, 6.4 pounds per car mile.

ENGINES.—Ball; three 150-h. p., single high-speed. Steam pressure, 95 to 100.

GENERATORS.—Six No. 32, belted direct from engines.

ROADBED.—Forty-five and fifty-two pound girder. Street length, 14 miles; length of all tracks, 15.5 miles. Seven miles track length paved. Company pave between tracks and maintain. In good condition.

GRADES.—Steepest, 10 per cent.; length, 775 feet. General character of road, moderate.

CARS.—Total number in equipment, 52; 25 motor (20 double, 5 single). Length, 16 feet closed, 21 feet,

30 feet, 32 feet open. All single track. Average number in use daily, 18 motor.

ANNUAL CAR MILEAGE, 699,060; daily average per car, 106.4.

WAGES.—Conductors and motor men, 15 cents per hour; engineers, $70 and $85 per month; mechanics, $2 to $2.25 per day; laborers, 15 cents per hour.

DETAILS OF OPERATING EXPENSES.

	Expenses per car mile. (Cents.)	
Maintenance of roadbed and track	.17	
Maintenance of line	.21	
Maintenance of power plant:		
Repairs on engines and boilers	.027	
Repairs on dynamos	.026	
Miscellaneous repairs	.037—	.09
Cost of power:		
Fuel (r. o. m. at $2.15 per ton)	.690	
Wages of engineers and firemen	.393	
Oil and waste	.058	
Water (at 10 cents per 1,000 gallons)	.054—	1.20
Maintenance of rolling stock:		
*Repairs on motors	2.086	
Repairs on car bodies and trucks	.239—	2.33
Transportation expenses:		
Wages of conductors and motor men	5.032	
*Cleaning and inspecting, and oil and waste for motors (see note below)	.....	
Miscellaneous	.035	5.07
General expenses:		
Salaries and office expenses	.544	
Advertising and printing	.086	
Insurance	.071—	.70
†Incidental expenses		1.23
Total		11.00

ROAD NUMBER SIX.

Length of road, four miles; grades, heavy; average number of cars in daily operation, five passen-

---

*In the item of repairs on motors are included gearing, cleaning and inspecting, oil and waste for motors, and also the expense incurred in changing the motors from closed to open cars in the spring and back again in the fall.

† Under this head are included a large number of small items which should probably be distributed about *pro rata* under the seven other heads.

ger, one freight; average daily mileage of passenger cars, 96.6.

DETAILS OF OPERATING EXPENSES.

	Expenses per car mile. (Cents.)
Maintenance of roadbed and track	.10
Maintenance of line	.06
Maintenance of power plant:	
Repairs on engines and boilers	.072
Repairs on dynamos	.017
Miscellaneous repairs	.007— .10
Cost of power:	
Fuel	2.147
Wages of engineers and firemen	.774
Wages of dynamo tenders and mechanics	.100
Oil and waste	.217
Water	.294
Other supplies	.232—3.76
Maintenance of rolling stock:	
Repairs on motors	.410
Repairs on car bodies and trucks	.176— .59
Transportation expenses:	
Wages of conductors and motor men	3.653
Wages of track sweepers, etc.	.237
Accidents to persons and property	.007
Cleaning and inspecting cars	.826
Oil, waste and other supplies	.134
Wages of freight hands	1.270—6.13
General expenses:	
Salaries of officers and clerks	1.370
Office expenses	.070
Advertising and printing	.261
Insurance	.168—1.87
Total	12.61

The above are results for a period of nine months. Coal cost $3 per ton for r. o. m. and $3.80 for lump. Wages of conductors and motor men, 9 to 11 cents per hour. Water cost 20 cents per 1,000 gallons. In the above no account is taken of the mileage of the freight cars. The cost of operation per car per day, including freight cars, is $10.11. Deducting cost of operating freight cars entirely, cost of passengers cars alone is 11.20 cents per car mile. In the nine months the freight cars hauled 16,738,000 pounds of freight.

This road is not one of the six already referred to.

ROAD NUMBER SEVEN.

Operating 18 miles by electricity and 4½ by horses.

Electric System.

BOILERS.—Babcock & Wilcox.

FUEL.—Three parts "pea and dust" (hard) to one part lump (soft). Average price, $2.23 per ton. Consumption, 4.7 pounds per car mile, costing .524 cent.

ENGINES.—One 400-h. p. Green, condensing, 50 to 55 pounds steam pressure, belted direct to three No. 32 Edison dynamos, and two 150-h. p. Corliss, condensing, 80 pounds steam pressure, belted to counttershaft, thence to three No. 32 Edison dynamos.

ROADBED.—Johnson girder rail, 45, 56 and 63 pounds. Length of all tracks, 18 miles; street lengths, 15 miles. All paved between tracks. Railroad company keep paving in order. Track in good condition.

GRADES.—Steepest, 8.4 per cent.; length, 500 feet. General character of road moderate.

CARS.—Thirty-six closed, double motors, 16 and 20 feet long; 25 open, single motors, 16 feet. Average number in use daily, 19 closed, 9 open.

CAR MILEAGE.—Total from April 1 to October 1, 1891, 549,177. Daily average per car, 107.

PASSENGERS CARRIED.—Total from April 1 to October 1, 1891, 3,508,168.

WAGES of conductors and motor men, $2 per day of 11 hours.

OPERATING EXPENSES, TAXES AND TOLLS.

	Expenses per car mile. (Cents.)
Maintenance of roadbed and track, based on experience of seven years at $900 per mile per annum	1.516
Maintenance of line, including material, wages of linemen and horse keeping	.569
Cost of power, including fuel, oil and waste, engineers and firemen, master mechanic, water, repairs to engines, boilers, station apparatus and building	1.304
Motor repairs, including material, wages superintendent and repair men, carpenter, blacksmith and machinist work	1.402
Wages of conductors and motor men	4.815
Sundry expenses, including starters, flagmen, foremen, watchmen, washing cars, sundry help, gas, coal, water, oil, repairs to car bodies and trucks by carpenters, blacksmiths and machinists	1.684
Extraordinary repairs, including repairing two burned cars, two dynamo armatures burned by lightning, $2,000 paid for exchange of 46 motor armatures, shaft repairs, and rent of extra station building	.725
Extraordinary horse expenses, including horses kept over and not used on horse road	.284
General officers, clerks and superintendents, less 20 per cent. charged to horse road	1.051
Legal expenses and damages, less 20 per cent., as above	.291
Insurance	.184
Property, dividend and capital taxes, and bridge rent and tolls	1.090
Total of all operating expenses, taxes and tolls	14.915

Depreciation.

The yearly depreciation on cost of boilers, electric lines, motors, and electrical and mechanical appliances connected therewith, is estimated at 10 per cent; the yearly depreciation on engines, dynamos, and appliances at 5 per cent. On this basis the total depreciation is 2.141 cents per car mile.

Horse System.

CAR MILEAGE.—Total from April 1 to October 1, 1891, 136,933, on 4½ miles of track.

OPERATING EXPENSES.

	Expenses per car mile. (Cents.)
Maintenance of roadbed and track, based on experience of seven years, at $500 per mile per annum	1.059
Power, including feed, shoeing, repairs to harness, depreciation of horses, barn men, etc	9.141
Conductors and drivers	7.896
Other items, including washing cars, foremen, watchmen, starters, gas, water, coal, oil, etc	1.464

Expenses per car mile.
(Cents.)

General officers, clerks, superintendents, etc., 20 per cent. of total amount.. 1.054
Legal expenses and damages, 20 per cent., as above................. .292
Insurance.. .179
All taxes and bridge tolls and rents................................ .980

Total of all operating expenses, taxes and tolls................. 21.975

NOTE.—In justice to the horse system, it might be added that for the past two years the price of feed has been much above the general average.

COMPARATIVE SHOWING.

	Electric.	Horse.
Receipts per car mile...	30.30	21.07
Operating expenses, taxes and tolls.........................	14.92	21.98
Net earnings..	15.38	

In an article on the cost of equipment, Mr. C. J. Field gives the following estimate:

I propose to take, as the best means of illustrating practically the purchase, equipment and operation of a street railway system with electricity, a city with a population of say 100,000—with a dilapidated street railway system, earning a gross income of $125,000, to purchase same for $500,000—property rights, franchises, etc.—and equip it with 40 miles of single track and 65 electric cars.

COST OF EQUIPMENT.

Steam plant (1,500-h. p. steam plant):

Five engines, 250-h. p. each, compound condensing, size 16 inches × 32 inches × 42 inches, with wheels weighing 30,000 pounds	$32,500	
Eight R. T. boilers, 72 inches × 16 feet	9,600	
Jet condensers	3,000	
Two boiler feed pumps...	900	
Steam and exhaust piping......................................	12,000	
Five engine foundations.......................................	3,500	
Eight boiler settings...	3,200	
Five 30-inch belts..	2,000	
Erecting and starting...	3,500	
Freight and miscellaneous.....................................	2,500	$72,700

Electrical plant:

Five generators, 200 kilowatts, $7,500.......................	$37,500	
Switchboard installation, foundations, etc..................	4,000	41,500

Building:

Power station, including stack, traveling crane, etc	$25,000	
Car house and repair shop, including tools, etc..............	15,000	40,000

Track construction:

40 miles girder rail construction, ties 2½ feet centres, 63-		
pound rail, etc., $1.15 per foot............................	$244,880	
Relaying, including paving, etc., at 60 cents per foot......	126,720	
Trucking, hauling, etc	24,000	
Ties, including 10 per cent. of joint ties, 130,000 at 40 cents	52,000	
Ties. including 10 per cent. of joint ties, 15,000 at 70 cents..	10,500	
		456,100

Line construction:

Ten miles iron poles, etc......................................	$75,000	
Ten miles wooden poles, etc...................................	40,000	
		115,000

Car equipment:

65 electrical equipments at $2,000	$130,000	
65 car bodies, 18-foot body, with open ends.................	65,000	
65 trucks at $250...	16,250	
		211,250

Summary:

Steam plant...	$72,700	
Electrical plant....................................	41,500	
Building	40,000	
Track	456,000	
Line construction	115,000	
Car equipment......................................	211,250	
		$936,550
Superintendent's and engineer's work...........	$50,000	
General and miscellaneous	50,000	
		100,000
		$1,035,550
Original purchase..............................		500,000
Total cost re-equipped..$1,535,550		

Gross income, say, $350,000.

Net income, say 35 per cent., equal to 8 per cent. on cost on the basis of an investment of about one million and a half of dollars, and from a property which in many instances was hardly earning its fixed charges formerly.

The cost of overhead construction he summarizes as follows:

Line construction per mile, complete, including track bond-ing, plain pole work, cross suspension or bracket with feed wire...	$2,000 to $2,500
With sawed and painted poles....................................	2,500 to 3,000
Iron poles, cross suspension, concrete setting, double track, feed and guard wires...	6,500 to 7,500
Same with centre poles..	4,500 to 5,500

He also appends a table, which will give a general summary of the cost of electric equipment of street railway systems, omitting the track construc-

tion, which, of course, varies with the number of
miles to be equipped.

COST OF ELECTRIC EQUIPMENTS FOR STREET RAILROADS.

No. of cars.	Steam plant h. p.	Capacity of generators. K. W.	Steam plant.*	Station electrical equipment.	Car equipments, car trucks and motors.	Line construction ½ mile of double track per car.	Total equipment (omitting track).
6	120	80	$7,000	$6,400	$19,500	$7,500	$40,400
10	225	150	11,000	10,500	32,500	12,500	66,500
15	375	240	17,500	15,000	48,750	30,000	111,250
20	450	300	22,000	17,500	65,000	40,000	144,500
30	675	450	28,000	22,000	97,500	90,000	237,500
50	1,125	750	50,000	33,000	162,500	187,500	433,000
100	2,025	1,350	90,000	60,000	325,000	375,000	850,000

The above figures are approximate only, and based on the best city railroad practice.

The cost of a single car equipped, including the
car body, truck and motors, he states, is from $3,000
to $3,500, and the cost of the electric part of the
power-generating plant is from $35 to $45 per horse-
power.

In a lecture by Professor Marks on high speed
inter-urban electric roads to be run at a speed of 2½
miles per minute, he states that a conduit could be
devised for containing the wires which would not
cost more than $35,000 a mile; the power, he says,
would not cost more than five cents per horse-power
per hour, including the *regie*.

The following table, taken from the U. S. Census
Reports, gives some statistics showing the distribution and costs of roads operated in different ways:

' Add 25 per cent. to these figures for Corliss.

Items.	All motive powers.	Distributed.			
		Animal.	Electric.	Cable.	Steam.
Length of line......	5,783.47	4,061.94	914.25	283.22	524.06
Length of all tracks.	8,123.02	5,661.44	1,261.97	488.31	711.30
No. of cars.	32,505	22,408	2,895	5,089	2,113
No. of employés...	70.764	44,314	6,619	11,873	8,158
No. of passengers..	2,023,010.202	1,227,756.815	134,905,994	373,492,708	286,854,685
Total cost.	$389,357,288.87	$195,121,682.50	$35,830,949.63	$76,340,618.23	$82,058,038.51

Commenting on this table, the *Street Railway Journal* stated: "Perhaps the most notable feature of the above table is the fact that in 1890 the railways operated by animal power were still far ahead of all others as regards their gross operating statistics. Rapid as was the advance of electric and cable roads the last five years of the decade, only a beginning was made in the supplanting of the older form of motive power. It is interesting also to see that although both in number and length of lines the electric railways have far outstripped the cable railways, the latter, nevertheless, represent twice as great an investment, operate nearly twice the number of cars and do a business more than twice as great. These relative figures point clearly to the far greater density of traffic upon the cable lines and to their large first cost. The cable lines, almost without exception, operate in the denser portions of the large cities, while the greater part of the electric roads are either suburban or serve the people of comparatively small cities."

Further tables, too large to be reproduced here, giving in detail the capital stock, dividends, interest, receipts and expenditures, taken from these census reports, will be found on page 25 of the January number, 1892, of the *Street Railway Journal.*

The table on the following page, compiled from various sources, gives some interesting data regarding some of the principal roads, and will explain itself.

As far as operating expenses per car mile are concerned, both cable and electric railways appear to be cheaper than those operated by animal power, as shown by the figures from the U. S. Census; the cable roads were built at a cost per mile of street occupied of over seven times as much as the electric railways were. The passenger traffic is about six times as great upon cable lines as upon electric railways. The figures correspond with the generally accepted fact that cable railways attain their greatest efficiency where an extremely heavy traffic is to be handled.

In the statistics for the six roads in Massachusetts, the cost of equipment was about half that of the cost of construction, and therefore about one-third of the total cost as given in the above table. On one of these roads the cost of repairing of bed and track amounted to $340 per mile, and on another $441. The repairs for cars and electrical equipment per car mile on one of the roads was .278 cent per car mile. The average car miles run per day on the six roads was 90.2.

The following table giving the receipts and operating expenses in cents per car mile, and the ratio

Measure	1	2	3	4	5	6	7	8	9
Per cent of operating expenses to receipts				74.4	55.0	70.0	73.0	85.2	78.8
Gross receipts in dollars				5,968,984	142,749	388,382	331,900	999,299	95,200
Ditto, in cents, per passenger	3.22	3.82	3.67	3.73			3.67	4.2	3.08
Ditto, in cents, per car mile	14.12	13.21	18.16	25.5	21.13	24.49	20.26	25.9	22.0
Cost of operating, in dollars	3,286,461	826,961	6,986,019	4,445,560	79,196	267,670	242,514	852,764	75,200
Car miles run				17,462,572	374,475	1,093,707	1,223,600	3,291,058	141,408
Total passengers carried	101,996,695	8,031,214	190,434,783	119,264,401			6,612,913	20,280,508	2,412,343
Total length of track in miles	143	67.22	552	260	81	179	18	106	12
Total cost, including equipment	$26,351,416	2,426,285	22,788,277	19,981,982				2,038,507	

Column key:

1. U. S. Census—10 cable roads
2. " 10 electric roads
3. " 30 horse car roads
4. West End horse (179 miles) and electric (81 miles): year ending Sept. 30, 1891
5. West End, April–August, 1891, electric lines
6. Ditto, Horse car lines
7. Pleasant Valley road, 1890
8. Six electric roads in Mass.; from Commissioners' report; 1 year
9. City and South London; first six months, 1891

between them, for a number of different roads, was compiled from various sources.

	Receipts in cents per car mile...	Operating expenses in cents per car mile...	Net earnings in cents per car mile...	Operating expenses in per cent. of receipts...
U. S. Census—10 cable roads; maximum.....	....	21.91	....	....
" " minimum.....	....	9.39	....	....
" " mean......	....	14.12	....	....
" 10 electric roads; maximum..	....	36.04	....	....
" " minimum ..	....	8.34	....	....
" " mean......	....	13.21	....	....
" 30 horse car lines; maximum..	....	27.02	....	....
" " minimum ..	....	9.10	....	....
" " mean......	....	18.16	....	....
West End line; April to August, 1891; electric	38.5	21.13	17.4	55.0
" " horse ..	35.0	24.49	10.5	70.0
Pleasant Valley road, 1890....................	27.55	20.26	7.3	73.0
Six electric roads in Mass., one year.........	30.4	25.9	4.5	85.2
City & South London line; first 6 mos. 1891..	22.4	17.6	4.8	78.8
Rochester line, June, 1891; electric..........	22.77	11.07	11.7	49.0
" " horse............	14.37	11.06	3.3	77.0
Birmingham (Eng.), 12 months; electric accumulator.....................................	40.5	19.5	21.0	48.0
Birmingham (Eng.), 12 months; steam......	31.5	22.0	9.5	70.0
" " horse......	21.5	19.0	2.5	89.0
" " cable.......	37.0	12.0	25.0	32.5
St. Paul & Minn., 11 lines, July..............	....	....	....	49.0
St. Paul & Minn., lines of heaviest traffic (9¼ miles)....................................	....	....	....	35.0
Chicago cable lines.............................	....	9.65	....	....
" horse "	....	21.9	....	....
Barking (Eng.) accumulator line, 12 months.	....	16.4	....	....
Budapest conduit line, 1890..................	24.0	....	....	....
" " 1891, August........	....	....	....	37.0

The following table gives the cost of operating per passenger carried:

Six roads in Massachusetts........................ 4.2 cents per passenger.
10 cable roads in U. S........................... 3.22   "   "
10 electric roads in U. S......................... 3.82   "   "
30 horse car roads in U. S....................... 3.67   "   "
City & South London road (6 months)............. 3.08   "   "
Pleasant Valley (Pittsburgh)..................... 3.67   "   "

In order to show the sub-division of the expenses per car mile the following table has been calculated from information obtained from various sources. Two of the roads are of the West End Line, Boston, one an electric and the other a horse car line.   It will be noticed that the expenses for car repairs are about two and a half times greater in the case of the electric line; but this difference is much more than made up in the saving in cost of motive power. The total cost per car mile for the Rochester line is only one half of that for the West End roads.   It should be ˙understood, however, that a strict comparison cannot be made, as there is a difference of opinion among those who give the figures as to what heading various expenses should come under. The figures given must, therefore, be understood to be only approximate.   For further data on this subject, see the article by Mr. Badger at the beginning of this chapter.

CENTS PER CAR MILE.	West End Line (Boston). For April to August (incl.) 1891. Electric trolley lines.	Ditto Horse car lines.	Pleasant Valley Line (Pittsburg). For the year 1890. Electric trolley line.	Rochester Ry. Co. June.	Barking Road (London). For the year 1890. Accumulator line.
Motive power.....	7.44=35%	10.74=44%	1.54= 8%	2.40=21.7%	5.7=35%
Conductors a n d drivers..........	7.14=34%	8.22=34%	6.80=34%	5.66=51.2%	2.9=18%
Car repairs........	1.32= 6%	.56= 2 %	1.68= 8 %	1. = 9.0%	6.6=40%
Other expenses...	5.23	4.97	10.24	2.01	1.2
Total cost per car mile.............	21.13 cents.	24.49 cents.	20.26 cents.	11.07	16.4 cents.

The following table gives the subdivision of the expenses per car mile for the Pleasant Valley Railway, of Pittsburgh, for the year 1890, and is doubtless quite reliable:

Conductors and motor men	per mile, 6.80 cts.	=33.5%
Motor and electric repairs	" 1.68 "	= 8.3%
Mechanical repairs	" 1.14 "	
Motive power	" 1.54 "	= 7.6%
Overhead system	" 0.45 "	
Maintenance of way	" 1.08 "	= 5.3%
General expense	" 0.88 "	
Stables	" 0.46 "	
Officers and salaries	" 0.84 "	
Interests	" 2.71 "	
Tolls	" 0.25 "	
General labor	" 2.43 "	
Total cost per car mile	20.26 cts.	

The following table gives some further data of interest for the same road, and is taken from the annual report:

EARNINGS.

Gross receipts for the year	$331,900.80

Total passengers carried during year, 6,612,913.

EXPENDITURES.

Pay rolls	$141,207.71	
Motor and car supplies	13,688.35	
Fuel and light	8,628.45	
Engines and boilers	2,767.69	
Dynamos	1,724.27	
Overhead system	1,713.33	
Roadway and stables	12,515.72	
General expense, taxes, bridge tolls, etc	33,268.12	
Interest on bonds	27,000.00	
		$242,513.64
Net earnings		$89,387.16
Surplus Jan 1, 1890		21,026.70
Total		$110,413.86
Paid dividend No. 31, July, 1890		39,000.00
Surplus Jan. 1, 1891		$71,413.86

The subdivision of operating expenses of the West End electric and horse lines (Boston) for one month (June, 1891) is shown in the following table; it should be remembered that the mileage of the horse lines was far greater than that of the electric lines:

Operating Expenses.	Electric.	Horse
General..	$7,465	$-2.217
Maintenance of track............................	3.334	9,923
Maintenance of buildings.........................	462	2,057
Maintenance of cars, etc..........................	4,281	6,579
Superintendence.................................	2,397	9,22 ;
Road and snow..................................	1,463	4,355
Injuries and damages............................	560	1.637
Car and lamp cleaners...........................	1,000	3,045
Conductors and drivers..........................	26,132	88,573
Total motive power..............................	$26,359	$116,210
Total operating expenses.........................	$73,459	$263,825
Net earnings...................................	$80,529	$131,729

A report of the Barking Accumulator Road (London) for the year 1890 contains the following figures showing the subdivision of expenses:

Wages at generating station...................	$3,280, or 3.3 cts. per car mile.	
Wages of drivers..............................	2,910, or 2.9 " "	
Fuel...	2,100, or 2.1 " "	
Oil..	310, or .3 " "	
Battery depreciation and repairs..............	5,730, or 5.7 " "	
Motor " "	860, or .9 " "	
Other " "	1.210, or 1.2 " "	
Total..	$16,400, or 16.4 cts. per car mile.	

The Birmingham (England) Central Tramway Company employs steam, cable, horse, and electric systems for operating its cars. The table on p. 132, taken from their report for the year ending June 30, 1891, will afford a very good opportunity to make comparisons of these systems. The city has over 400,000 inhabitants, and it is stated that the people use the cars very freely. It is also claimed that mechanical traction has been employed to a greater extent in that city than in any other in Great Britian and Ireland.

Commenting on this table, the *Street Railway Journal* states: "A great preponderance will be observed in the steam department over the other three. This is, of course, due to its more extensive

EXPENSE ACCOUNT.

	Steam Department.	Horse Department.	Cable Department.	Electric Department.
FOR MOTIVE POWER.				
Wages.........................	$49,326.20	$31,358.28	$15,395.40	$7,268.90
Fuel.........................	45,289.34	....	6,997.26	4,625.38
Forage and bedding........	....	46,775.26	....	....
Water and gas..............	5,174.62	804.76	737.34	190.10
Veterinary and shoeing....	....	5,094.00	....	....
Harness repairs.............	....	2,053.60	....	....
Stable utensils..............	....	563.72	....	....
Stores........................	8,328.08	....	1,932.66	2,032.90
Sundries.....................	1,644.80	954.80	299.46	260.52
Repairs—Wages.............	21,269.12	....	772.90	5.32
Materials.....................	21,253.18	....	9,284.48	828.62
Renewals....................	....	5,570.60	....	....
Totals....................	$152,285.34	$93,175 02	$35,419.50	$15,211.74
CAR REPAIRS.				
Wages......................	$4,256.08	$3,640.74	$1,774.56	$1,689.56
Materials....................	3,542.68	3,263.16	6,991.38	3,694.94
Totals....................	$7,798.76	$6,903.90	$8,765.94	$5,384.50
TRAFFIC EXPENSES.				
Wages	$32,267.70	$13,549.64	$11,100.14	$3,000.62
Water and gas..............	2,719.04	442.00	737.32	190.14
Stores......................	1,368.00	430.84	446.02	70.66
Stationery, tickets, etc......	3,182.22	1,289.80	1,113.56	361.54
Sundries....................	604.22	553.82	254.66	114.60
Totals....................	$40,141.18	$16,266.10	$13,651.70	$3,737.56
PERMANENT WAY AND BUILDINGS.				
Wages	$6,310.80	$212.28	589.48	$207.70
Materials....................	30,668 04	1,542.20	1,243.42	172.64
Totals....................	$36,978.84	$1,754.48	$1,832.90	$380.34
TO GENERAL CHARGES.				
Stationery and incidentals..	$1,815.42	$762.40	$510.62	$195.98
Salaries	2,381.16	1,249.10	1,040.58	279.82
Compensation...............	3,870.14	363.92	155.44	213.56
Rates, taxes and insurance	8,626.56	2,551.08	3,769.50	1,423.06
Professional charges.......	7,258.66	1,010.34	678.82	528.12
Sundries....................	1,263.92	798.62	698.46	309.12
Totals....................	$25,215.86	$6,735.46	$6,853.42	$2,949.66
No. miles run..............	1,184,401	131,528* 506,196†	522,876	138,396
Passengers carried........	14,242,827	1,114,388* 2,638,028†	5,241,362	1,144,718

* Street cars.
† Omnibuses.

use over the cable, horse, or electric systems. The electric department makes a very good showing, but is defective, as will be explained later. The cable division, in our opinion, makes a much better showing than its associates, considering the fact that but twenty 14-foot cars are operated in this department. The horse department is almost monopolized by the 'buses, but the cars make a good showing. On the whole, the report appears satisfactory. The net profits, after deducting the expenses from the gross receipts, show almost 4¾d. or about nine and a half cents, in the steam department, 1¼d. in the horse department, 12½d. in the cable department, and 10½d. in the electric department. The cost to operate the steam division was twenty-two cents, the horse division nineteen cents, the cable division twelve cents, and the electric division nineteen and a half cents. However, for a city of the size of Birmingham more than 24,381,323 passengers should have been carried in one year, especially since rapid transit is employed here to a greater extent than in any other city in Europe. The cable division of the Birmingham company has been a most pronounced success, as is always the case with the cable system where it is given a fair trial. The writer was informed, by one of the company's engineers, that the net profits were already very large, and to materially increase them the fares were to be reduced, and a great increase for the future was confidently expected. Twenty 14-foot cars, mounted on bogie trucks, are run by cable power, and considering that they carried over 5,000,000 passengers during the last year they may be deemed most sat-

isfactory. It is surprising, however, that the profits
are so large, considering the service given. The
cars are very dirty, and we are quite sure that they
would not be tolerated in America. The conductors
and gripmen, or drivers, as they are called in Birmingham, look quite as shabby as the cars. They
stand in strong contrast to the well-uniformed men
in the cable service at St. Louis, San Francisco, and
New York. The steam and electric cars are much
cleaner, and as it is with the latter we have to deal,
we shall proceed to do so immediately.

" The electrical division using storage batteries was
opened on July 16, 1890, and inspected by the Board
of Trade July 24 of the same year. The latter
approved it, and from the very first it has been well
received by the public, and so much so that the running of additional cars is contemplated. But let it
be understood from the first that the road is an
experimental one, like most storage battery roads
now in existence, and it is upon the economical
running of the road in the future more than upon
its past record that its adoption depends. We were
told by a person who is an authority, that the road
was reported to have earned $15,000 clear profit, but
in reality there was a deficit of $5,000, because the
storage battery account was not handed in, and, as
a great many no doubt know, this is one of the
greatest items of expense attending a storage battery road."

A report on the two-mile open conduit at Black-
pool, England, gives the following figures. There
are 12 cars in operation, which have run 98,000 car
miles and carried 9,034 passengers during the year,

which is an increase of 6 and 15 per cent. respectively. The dividend declared was 7½ per cent. The conduit repairs for one year amounted to £144, or £72 per mile. The motor and car repair amounted to £192, or £16 per car, which is about five per cent. of their original total cost, and about 13½ per cent. of the original cost of the complete electrical apparatus.

A report on the Rochester railroad for the middle of this year states that they operated 44 vestibule electric cars 18 feet long; that the gross receipts were 23.05 cents per car mile for a mileage of 159,-567. The cost of operation of these cars for one month was $18,332, and the receipts were $37,053, leaving a net profit of $18,721. The division of expenses will be found in one of the above tables.

The average cost of haulage in London for street cars and omnibuses is said to be about 20 cents per car mile, while the cost of horse haulage in that city is given as 12 cents per car mile.

In an article by Mr. Field, he states that the cost of generating power for 16 and 18-foot cars is from three to five cents per car mile for expense at generating station. For 33-foot cars, or trailers, he assumes that the cost will be 50 per cent. more.

In a report of some figures from six electric street railways in different parts of various States, in which the Thomson-Houston system is used, it is stated that the average amount of coal used per horse-power hour at the power station is about five pounds; while the average amount used per car mile on the road, computed from a mileage of over 1,200,000 miles, is about eight pounds, or over 50 per

cent. more than the first quantity. It is also found from the figures obtained from these roads that the average cost of coal at the power station is $3 per ton. From this it is seen that the cost of fuel per horse-power hour at the station and per car mile on the road are respectively .75 and 1.02 cents.

CHAPTER V.

OVERHEAD WIRE SURFACE RAILWAYS.

Under this heading will be included what is usually known as the overhead trolley system. The general construction is so well known that no description need be given here. In general the current is led from the power station to different portions of the line by feeders, which are in some cases underground, but more generally overhead. The feeders are connected at various points with the trolley wire, which is usually in separate sections. From this the current is taken by the trolley on the roof of the car, passes through the motors, thence to the iron framework of the car, and from there to the track, returning to the station either through the track or through the ground.

By far the largest number of roads are run by this system, and it may be said that this is the only one which is at present in extended use. Its success is already far beyond dispute, as the rapid introduction of the system shows.

In many cities objections have been raised to this system, chiefly by the municipal authorities. What these objections are, and whether they are valid or

not is best seen by the answers which were published by a number of writers, of which we give
here a few of the most important.

Regarding the overhead system, Mr. F. H. Monks
writes as follows: "The only system for the operation of street cars by electricity which has, up to
date, met with commercial success is the overhead
wire system, and, therefore, the application of electricity as a motive force at present must be confined
practically to that system. The extension of the
overhead system in the large cities of the country
has undeniably been greatly checked by the opposition of the municipal authorities to the erection of
poles, the stringing of wires, the fear that loss of
life or injury to persons would ensue as an inevita-
ble result of the operation of cars by electricity, and
the belief that any line so operated would of necessity succumb to the rigors of a Northern winter.
Notwithstanding such opposition, the number of
roads operated by the overhead wire system has
increased with great rapidity, and after the people
have had an opportunity to learn to their complete
satisfaction that the poles and wires, though certainly objectionable, are yet justifiable under the
circumstances, that absolutely no one is killed by
the electrical current and that no snow storm, however severe, has any terrors for the managers of the
line, they clamor for the rapid extension of the system, as they certainly are doing in Boston to-day.
The progress of invention respecting the development of the overhead system has been marvelous,
and the great danger has been the possibility of
getting too far committed to present methods and

appliances to prevent the acquirement of the improved devices, which are coming into sight almost daily. These facts have unquestionably acted as a deterrent with street railway managers everywhere to the even more rapid extension of the overhead system. I make no doubt but that within a few months matters of detail respecting this system will be so generally regarded as settled that its extension will be more rapid than in the past."

An editorial in the New York *Tribune*, referring to the overhead trolley system, says: "It is dangerous, inefficient, destructive, and unsightly. That the overhead wires are destructive, firemen have repeatedly declared, and proven to the dismay of numerous owners and tenants of burning buildings. Under perfectly favorable conditions, it does undoubtedly do its work reasonably well, but it is peculiarly liable to break down just when the public need of it is greatest. An ice storm is always expected to interrupt its course, even if it does not bring the wires in a tangle to the ground. The experience of Boston in this particular has been exceptionally uncomfortable and irritating, but the same inherent defect has been experienced elsewhere. Lastly, the existence of overhead wires carrying a current of electricity capable of operating a line of street cars is, and in the end of the case must be, a constant menace to public safety and a common cause of fatalities. In cities the trolley system is intolerable."

To this editorial, Mr. S. J. MacFarren published a reply, of which the following abstract is taken. He states that recently printed replies from the mayors of nearly 60 cities and towns possessing electric traction showed that nine-tenths praised the trolley rapid transit system with more or less enthusiasm, and not one condemned it. (See *Electrical World*, Nov., 1890). Many of these officials denied explicitly that the trolley system is considered dangerous by their people. He states that electric railway wires have never killed or maimed a human being. He claims that the firemen's complaint against overhead wires was generally, if not invariably, against light and telephone wires, which are strung by the hundreds over the sidewalks where fire ladders are to be hoisted, while electric railway wires need be but two, and those in the middle of the street. Regarding the stoppages in storms, he claims that the electric railway motor is immensely superior to animal traction in this emergency, and when properly equipped and managed, is equal to the cable.

Regarding the question of the danger of the overhead system, Mr. F. L. Pope states: "With a view of getting at the actual facts in the case, the *Boston Advertiser*, a few months ago, sent out a circular letter asking for information from every city in which electric railways are in actual operation, from Portland, Me., to Galveston, Tex. It was asked what system was used in each place; whether there had ever been loss of life or injury from the wires; whether there was any serious objection on the part of the public to overhead wires, and what was the general opinion in the locality as to the effect of the introduction of electricity upon the street railway service. Replies were published from 64 cities and towns. All but four of them were favorable. Not one solitary instance of accident or serious injury from electric currents was reported. One of the objecting places was Newport, R. I., where it seems the "upper ten" strenuously opposed the introduction of anything that would popularize riding on the streets.

"Many persons are alarmed at the vivid flashes of light which are often seen at night beneath the wheels of an electric car, and at the point of contact of the trolley wheel with the overhead wire, and are under the impression that they must indicate a very dangerous electric pressure. Such, however, is not the case. In an electro-plating establishment at Ansonia, Conn., I once saw a workman accidentally set a tin pail filled with water upon a pair of electric conductors near the dynamo. The pail instantly disappeared, being not merely melted, but being converted into metallic vapor, with a terrific flash which illuminated the whole building with a dazzling and instantaneous radiance; yet the current which produced this startling phenomenon was of such low pressure that it was impossible to detect its presence by the sense of touch, even by applying the hands directly to the conductors.

"An extraordinary amount of nonsense has been printed and talked in respect to the alleged dangers of both electric light and railway wires. The public have been needlessly alarmed by the exaggerated statements of interested parties, but, nevertheless, the danger is so small, as a matter of fact, that the actual figures are almost astonishing. Most of the accidents which have been reported occurred in New York city. Yet the statistics show that in

1889, out of 1,467 deaths in New York city by accidents of various sorts, only nine were due to electricity, a considerably less number than were killed by being run over by horse cars. Not a single death was recorded in Boston, although there are perhaps more wires there in proportion to the population than in New York. There are in the six New England States nearly 140 arc light stations, burning over 20,000 arc lamps and distributing 30,000 horsepower of electricity through the streets of the principal cities and towns. During the last ten years, so far as I can ascertain, there have been but five deaths from electricity; four of these were employés of the lighting companies, and one was a careless boy who climbed upon a shed and took hold of a wire. During the same ten years the steam railroads of New England have killed 2,339 employés and 2,902 other persons; 5,241 against 5. Not only is electrical power far less dangerous than the same quantity of power used in other industries, but it is relatively safer, as the few accidents that do occur are among the employés. These remarks refer to electric wires in general. Now as to electric railway wires:

"I believe that it is an incontestable fact that not one single man, woman, or child has ever been killed or even seriously injured by a 500-volt current, which is the highest pressure ever permitted upon electric railway wires. Every alleged case of accident by railway wires has, upon investigation, proved to be either without foundation, or to have been caused by an electric light current. When we consider that shocks have been experienced by men, women, and children, persons of all ages and all sorts of physical condition, sometimes for a period of several minutes, experience seems to warrant the positive assertion that the electric railway current is not dangerous to human life, and that we may dismiss that question from further consideration."

On the same subject Mr. Griffin states: "On this point I think the public are now well satisfied. While there are few employés on any of the roads now in operation who have not had the full shock of 500 volts repeatedly, there is not a single instance of any of the patrons of these roads who have been killed or even seriously injured by the 500-volt current from the overhead wire. Electric cars will run over and kill the careless pedestrian or the drunken passenger who falls from the platform in front of the wheels, as will the horse car, but no passenger or pedestrian has ever been killed by the trolley wire, and statistics do not show that the electric car is in any respect any more dangerous to life than the horse car or cable car. Last year (1890) the West End street railway system of Boston carried 114,853,081 passengers, and all the steam railroads of the whole State of Massachusetts only carried 98,843,712. The West End system killed 15 passengers and employés, and the steam roads killed 325. Of the 15 fatal accidents on the West End system, five were attributed to electric cars and 10 to horse cars. It is only fair to say that the narrow and crooked streets of Boston and the enormous traffic of the West End system are conditions peculiarly conducive to accidents.

"In the year 1889 nine human beings were killed by the arc light wires in New York city (2,500 volts), and the authorities were roused to such a pitch of frenzy that the poles were chopped down and a large part of the city left in darkness. Yet, with perhaps one exception, all of the victims were employés of the lighting companies, and suffered because of failure to observe proper and well-known precautions. In the same year twelve persons were asphyxiated by gas and over thirty were killed by signs and other objects falling on their heads as they walked peacefully along the streets. In time we are able to estimate every danger relatively, but in the beginning, unknown dangers, those to which we are not accustomed, are greatly exaggerated."

In a compilation of data obtained from a number of companies, in answer to a set of questions, Mr. Mansfield states: "I am happy to state that under this heading not one road reports as killed or even seriously injured an employé or passenger by the electric current, or falling trolley or span wire. Several report employés as receiving shocks, and one of a boy throwing a wire over the trolley wire and receiving the full potential of the current. None, however, were seriously injured. Several accidents are reported of collision and running over, but these cannot be entirely avoided, and are inherent in any system."

Descriptive.

A very fair idea of some of the larger overhead wire roads in operation may be had from the following extracts, taken from the published descriptions of some of the largest and most typical plants. Owing to the difficulty of getting full and reliable descriptions, some of these are necessarily very incomplete.

West End Railway, Boston.—This line comprises the street railways of the city of Boston, and is run partially by horse and partially by electricity, using the overhead system. It is the largest electric railway plant in the world. The mileage of the electric system was about one-quarter of the whole mileage. On September 30, 1891, they owned 244.47 miles of track and leased 14.63 miles, a total of about 260 miles; of this 81.234 is equipped with electric overhead system, 18.907 is partially equipped. The average receipts per passenger were for both horse and electric lines 4.938 cents. Total mileage of the electric cars was 4,588,146 car miles, or 26.27 per cent. of the whole, while the horse cars ran 12,874,426 miles, or 73.73 per cent. They contemplate converting the whole road into an electric one as soon as possible, and to get the cost of transportation down to 16 or 17 cents a car mile. The cost of running in the dense parts of the city is considerably greater on account of the slow speed. In these parts they estimate that the average speed is from one to two miles an hour. Further statistics will be found tabulated, together with those of other roads, under the heading of cost of construction and operation.

Minneapolis Street Railway.—This is claimed to be one of the most complete and extensive systems in operation. The novelty of the system employed throughout lies in the fact that there is not a wire in sight in the heart of the city except the overhead trolley wire. The feeders, mains and track feeders are contained in a conduit underground, the trolley wire connecting with the feeders by means of a sub-feeder through the hollow iron supporting poles. The conduit is located between · the tracks, and is built as follows: A two-inch plank, first treated by boiling in fernoline, is used for constructing a long trough of the desired size. This trough is so nailed together as to be continuous, and without joints from manhole to manhole, a distance of 408 feet. The trough is placed below the surface at such a depth that the top is six inches below the paving blocks. The conduit proper consists of a number of heavy paper tubes of the Interior Conduit and Insulation Company's make. The tubes employed are one inch or one inch and a quarter inside diameter, laid in the trough in ten-foot lengths, and separated from each other and the sides and the bottom of the trough by rings or spacers. The tubes are made continuous from manhole to manhole by use of a telescopic joint. After the tubes have been put in place, pitch is poured in, filling the interstices, and leaving the tubes with a solid insulating filling, impervious to moisture, around them.

It is claimed that the system is the first installation of underground conductors ever made in which copper wires were drawn into a conduit without other insulation than the conduit itself. There is at the present time about 60 miles of bare copper cable resting in the conduits, varying in size from 100,000 to 500,000 centimetres. The insulation resistance on the entire amount of tubing with overhead trolley and outlying feeders, as shown by a test, is 1,081,-

147 ohms. Some recent tests of the feeders in the conduits show an insulation resistance as follows: Feeder, a, 37,719,598 ohms; b, 18,647,000; c, 2,251,-000; d, 10,298,000; e, 1,791,000; f, 1,815,000; g, 1,488,-000.

The population of Minneapolis and St. Paul is 350,000. There is not a horse car in either city. Minneapolis has 120 miles, with posts in the middle of the streets, with arms extending over the track on either side. St. Paul has 90 miles, 75 of which are electric and 15 cable. The cable will be kept only for a portion which has a grade of 17 per cent. They use Lima oil for fuel, at a cost of $1 a barrel. Three barrels are said to be the equivalent of a ton of coal. It is said to cost them $1 a day for power for an electric car for a system of 150 cars; horses used to cost them $3.85 to $4 a day per car, even with the low priced grain. The principal item of repairs is the burning out of the armatures. The line of their heaviest travel, $9\frac{1}{2}$ miles long, between Minneapolis and St. Paul, shows a cost of operating of only 35 per cent. of the receipts.

Federal Street and Pleasant Valley Passenger Railway (Pittsburgh).—Mr. D. F. Henry gives the following description, presumably for the year 1890: "There are about 58,000 feet, or 11 miles, of track laid with the improved Johnson girder rail; 31,000 feet, or about six miles, were relaid and completed with tram rail; there are now about 18 miles of single track. The large substantial fireproof power station is equipped with two Hazelton tripod boilers, of 500 horse-power each; also a battery of two hori-

zontal tubular boilers, of 125 horse-power each;
total, 1,250 horse-power, with the Roney patent
smokeless stokers, tanks, coal and ash conveyers,
steam pumps, heaters, and purifiers; an artesian
water well, with steam pump; two large tanks for
a water reserve in case of accident, or shutting off
the city water supply; also a fire protection. The
coal bins have a capacity of 1,000 tons, or 40 days'
supply.

"There are three Buckeye engines, two of 400
horse-power each, and one of 200 horse-power; total,
1,000 horse-power; five Edison dynamos, of 80,000
watts each, or combined, 535 electric horse-power;
three Thomson-Houston dynamos, of 60,000 watts
each, or combined 240 electric horse-power, making
a total of 775 electric horse-power; all connected
with main shaft, with idle pulleys thereon to attach
two additional dynamos of 80,000 watts each, which
would balance the engine capacity. There would be
still in reserve 250 horse-power boiler capacity.
They have contracted for a 250 horse-power engine,
also dynamo of equal capacity. They will have
ample power, with a reserve both in steam and electricity, for all requirements of the near future.   In
the power station there still remains room enough
for considerable power for future extensions.

"They have remodeled the buildings into repair
shops, with suitable examining pits; machine shops,
with engine, elevator, lathes, drill presses, planers,
milling and boring machines, emery and grinding
wheels, and blacksmith forges; electric shops, with
lathes, dryers, and insulating apparatus for the
building of new car motors, and station equipments

and their repair; a carpenter or car shop, with engine, planers, lathes, saws, boring machines, for building new or altering cars suitable for electric service; a complete stable outfit, with first-class horses, harness, wagons for hauling coal and other supplies, track sweeper, water carts, etc.

"The car house covers a space of 140 by 145 feet, having a capacity for 75 cars. This will afford ample storage room for all cars on the main line, as well as all not in daily use on all the other divisions.

"Upon the completion of the branches they expect to have one of the most extensive and valuable street railway properties in the country, as well as one of the most complete electric roads in all its details in the world. They have now 50 motor cars, 25 trail cars, 6 snow plows, 5 salt cars, 1 snow sweeper, a stone crusher, engine and boiler, enabling them to ballast the suburban roads at the least possible expense. They claim to have one of the most difficult roads to operate in existence, with streets having over one hundred curves, many of which have a short radius, and with heavy grades, leaving but little straight and level roadway.

"In the twelve months that the road has been in operation, the total time lost was but 20 hours, 3 of which was on account of the shutting off of the water supply to the boilers, and 3 hours from engine and line troubles. The other 12 hours was caused by two of the most severe snow storms experienced for many years. The first snow fall was 16 inches and the last 10 inches. With the narrow streets and sharp curves, the sweeper piled the snow on either side of the track to the height of from 4 to

6 feet; consequently the snow fell back upon the track, thereby destroying the traction and derailing the cars, but not, as has been erroneously reported, interfering with the electric contact. Steam, cable, or even horse cars, under the same circumstances would have been compelled, as we were, to resort to shovels. The wires, too, during this storm gave no trouble; but they were somewhat annoyed by the falling of telegraph wires upon theirs, which were easily and quickly removed.

"They have been in operation nearly a year, and during that time not one single person has received the slightest injury while on board their cars, nor has any person or animal been injured by their wires upon the streets. I believe this record to be unsurpassed by any company handling a like number of passengers, with either steam, electric, cable, or horse power."

Some statistics about this line will be found under that heading.

The Riverside Park Electric Railway.—This plant, in Sioux City, Ia., which was opened in June, is stated to be in many respects a model plant. The following description is therefore given here in full: The power house is located at the edge of the park near the confluence of the Big Sioux with the Missouri River. The station is a solidly constructed brick building 62x95 feet, divided into two rooms with fire wall between. One room contains two Wicks tubular steel boilers, 60 inches by 16 feet, having forty-four 4-inch tubes; the base of stack, which is 113 feet in height, is 42 inches in diameter, and built of No. 10 and 12 iron; the Worthington pump; a 250 horse-power Kroeschell heater, two Hancock inspirators, and four sets of steam loops. The second room contains the two 125 horse-power compound Westinghouse engines, at 100 pounds steam pressure, having special flywheels of double the ordinary weight, insuring close regulation under the sudden changes incident to heavy loads. Coupled directly to the engines by two of Schieren's perforated electric belts, 16 inches in width by 53 feet long, are two Westinghouse compound wound generators of the improved U. S. type, each of 100 horse-power capacity, having 54-inch paper pulleys running at 350 revolutions. These generators rest on a brick foundation 6 feet in depth, and leveled off with a half-inch layer of melted sulphur, on top of which rests a hardwood base 8 inches in height, that is covered with three sheets of one-eighth inch asbestos previously saturated with P. & B. paint, and thoroughly dried. On this the rails are placed and firmly fastened to the foundation by insulated bolts. Powell's patent oil cups and lubricators are used on both engines and generators. The dynamo leads are carried from the generator through rubber tubing down to and under the flooring and to the switch-board of polished sycamore with marble face on which is placed the improved form of Westinghouse voltmeter, lightning arrester, automatic cut-outs, switches, etc. An independent circuit is used for the light in the station and car house. The car house is within 100 feet of the station and is a brick building 50x150 feet, designed to hold 20 cars, and containing every requisite for the rapid handling of the same. The present equipment consists of six new motor cars, eight old trailers, and the steam dummies. Each of the new cars is equipped with two McGuire improved trucks, each of which supports a 30 horse-power Westinghouse single reduction motor, especially built for heavy work required on this line. The nine miles of roadbed is nearly a continuous series of curves, there being 38 in number, ranging from 4 degrees to 20 degrees, and has been built through several deep cuts, from one of which over 60,000 cubic yards of earth were removed, at an expense of $4,800, while close to the power house there is a trestle built 634 feet long, having a rise of 4 per cent. on a 19-degree curve of 300 feet, the upper end of the trestle being 16 feet above grade. The pole-line circuit is a model one, every pole being truly aligned and, when necessary, properly guyed. The feed wires are Roebling's weather-proof 000, and the trolley wire No. 0, bare copper, suspended by means of a special form of insulator attached to a bracket arm bolted to the pole.

The Sissach-Gelterkinden Electric Street Railway. —Although America is in the lead in the electric railroad industry, we may often learn something from the careful and intelligent workers in this field abroad. In this connection a few details regarding the construction of this small Swiss railway, opened in April, and some of the results of its operation, may be of interest. From a recent article on the subject in *Schweizerische Bauzeitung* by Dr. A. Denzler, it appears that the whole length of the route is 3.25 kilometres, starting at the central railway station in Sissach and running by a somewhat circuitous route to the terminus at Gelterkinden.

The maximum grade on the whole line is 15 per cent., and the sharpest curves are two of 60 metres radius, one of these being on a 12 per cent. grade; the gauge is one metre.

The central station and turbine house are one kilometre from Sissach, the beginning of the line. The power is furnished by a Jonval turbine, which, with the mean quantity of water at a flow of 6.75 metres, is about 40 horse-power. The water is supplied by the river Ergalz, and is led to the turbine house in a channel partly open and partly closed. As the dynamo makes 600 revolutions per minute and the turbine runs at a speed of but 98 to 100 a spur gearing is interposed between the wheel and the dynamo shaft. A so-called brake regulator is used, by means of which the turbine always runs under full load, that part of the power which is at any time not absorbed by the dynamo being taken up by the brake of the regulator. This method has been employed quite extensively in this class of work abroad, but of course it can be economically used only in those cases where the quantity of water consumed need not be taken into consideration.

The dynamo which furnishes the current for the car motors was built for a normal output of 50 ampères at 700 volts pressure, or, in other words, about 35 kilowatts. It is a two-pole machine with series windings and a flat ring armature. The current is taken from the bronze commutator by two pairs of brushes of copper wire gauze. It is said that by using this style of brush instead of those of sheet copper the disturbing effects of induction are not felt so much in the telephone circuits in the neighborhood of the railway lines.

The foundation of the dynamo is insulated from the ground by wooden beams, the positive pole being connected with the ground and the rails which end at the station. On the switchboard are mounted a main circuit breaker, with carbon contact points, an ampère meter, a lightning arrester, and an automatic short circuiting apparatus. The first mentioned of these is designed for short circuiting the field magnet windings of the dynamo so that the current is interrupted as soon as the maximum potential has been reached. The discharge points of the lightning arrester consist of carbon rods so arranged that they can be easily replaced and carefully adjusted. Any arc which may be formed over these terminals by the dynamo current following to the ground when a lightning discharge has taken place is extinguished by an arrangement which consists of a solenoid in the earth circuit, the core of which is drawn down by the continuous passage of the current through its coils. By this means the two contact points established between them is extinguished. Special apparatus for regulation are not used, so that the required duties of the attending engineer in the station are considerably reduced.

The rails are connected by strips of sheet copper, which are soldered and bolted to the rails; when another type of rail is used, the connection is made by copper wires which are soldered to hooks of sheet copper bolted to the side of the rails. In addition to this the two rails of the track are always electrically connected by copper wires of six millimetres diameter at intervals of four rail lengths.

In addition to the overhead trolley line which forms the return, a feed wire is placed upon the bracket poles; this feed wire and the trolley line are connected in multiple arc by insulated cables at intervals of about 100 metres. By this means the overloading of the trolley wire is prevented and at the same time it is possible to use the current when the trolley wire at any section is broken.

The trolley wire consists of hard drawn copper wire six millimetres in diameter placed at the uniform height of $5\frac{1}{2}$ metres above the rails. The feed wires are run on oil insulators, while the trolley line insulators are of the Sprague type. The author of the article in question points out the fact that the awkward and clumsy liquid insulators form a striking contrast with the little bell insulators used on the trolley wires.

The trolley line insulators are fastened to plain brackets made of tubing. The brackets are secured to poles of round timber which are placed alternately on opposite sides of the street. The poles are 35 to 40 metres apart, and although it is not to be pretended that these poles are ornamental, it cannot be said that they disfigure the streets; on the contrary, one gets the impression that when light and well designed iron supports of a little greater height than those here described are used, overhead wires could be employed without hesitation in the broad streets found in most of the suburbs of Swiss cities.

The truck and the motor used will be found illustrated and described among the others, under those headings.

The current flows from the car wheel into the iron frame and then branches out into the field magnet coils of the two motors, which are connected in multiple arc; from the field magnet coils the current passes into the armatures, which are connected in series with the fields. After passing through the armatures the lines from the two machines join and pass through the regulating resistance and the ampère meter.

Between the armatures and the field magnet coils is placed a double two-pole reversing switch, by which the circuit may be entirely opened or the direction of the rotation of the armatures may be reversed. The speed of the train is regulated by resistance coils, which are arranged about a central spindle. The hand wheel by which the mechanical car brake is applied is mounted within easy reach of the motor man, and is directly above the handle of the rheostat switch. It is claimed that by operating both of these hand wheels at once the motor man can stop the train within 20 metres, even when running at full speed. An alarm bell is placed on top of the locomotive, the end of the bell cord hanging directly at the side of the motor man.

The total weight of the locomotive is 6,170 kilogrammes, the total length being 4.69 metres and the total height 2.94 metres. Usually the train consists of a locomotive and two passenger cars, one of which has a seating capacity of 24 and the other a seating capacity for 12 persons, besides considerable space for packages. The cars weigh 4,140 kilogrammes, and their total length is 7.8 metres. Dr. Denzler points out that the weight of the cars seems decidedly too great when it is considered that a common closed horse car for 24 passengers weighs only 1,300 to 1,500 kilogrammes. Even in the favorable case, when the train consists of three fully loaded cars, the paying load of 72 passengers is 5,400 kilogrammes, while the locomotive weighs 6,170 kilogrammes, the passengers cars 16,420 kilogrammes, the train crew 150 kilogrammes; the total load is 22,740 kilogrammes. From this it will be seen that the paying load is but 23.7 per cent. of the dead load, and that but 18.3 per cent. of the total weight transported brings any revenue to the company. Dr. Denzler says: "If we compare this with the general results of an American motor car for 24 passengers equipped with two motors of 15 horse-power each, we find the mean values for the corresponding quotients 47 per cent. and 36 per cent., that is about twice as much as those given above. Even in those cases where sufficient power is available it is of no consequence to increase the efficiency of the locomotive by improved construction when the advantage thus gained is again lost by the use of too heavy cars."

For lighting the train it was not thought best to use electric lamps, since the frequent changes of the coupling connections between the locomotive and the cars would extinguish the lights. It is proposed in the near future to use electric heaters throughout the train.

The tests made to ascertain the amount of power

consumed are of considerable interest. In Fig. 22 are shown the fluctuations of the output in watts as measured at the terminal of the dynamo at intervals of 15 seconds during a run of one kilometre. During this run the minimum and maximum output were 13.1 and 25.2 kilowatts respectively. The aver-

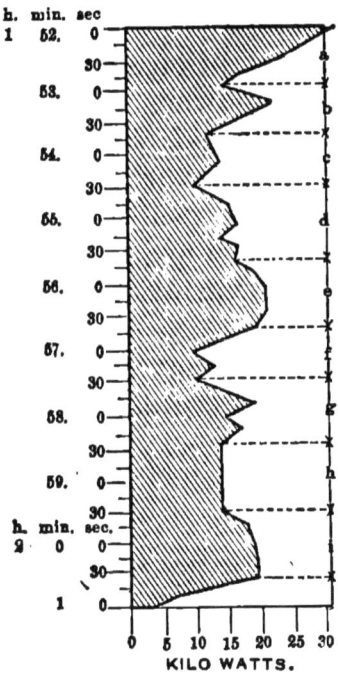

FIG. 22.—POWER CURVE.

age of 34 measurements was 16.7 kilowatts, so that the highest and lowest values are respectively 50.8 per cent. above and 21.5 per cent. below the mean value.

The potential dynamo terminals varied from 560 to 760 volts. The current usually varied during the run from 17 to 30 ampères, the maximum consumption being 50 ampères, this reading being taken at the moment of starting on a 15 per cent. grade, the total output at the same being 30.2 kilowatts. During this run the train consisted of the locomotive, 6,170 kilogrammes; three passenger cars, 12,420 kilogrammes; 30 passengers, including conductor and brakeman, 2,250 kilogrammes, or a total weight of 20,840 kilogrammes.

The largest consumption of power was 33.5 kilowatts, or 45.5 horse-power. This was observed at the starting of a train of 33 tons weight on a level track. Even here the variation from the mean value during the run was but 92 per cent., while with motor cars, which are provided with a double reduction gear, this variation is often more than 200 or 300 per cent.

An idea of the efficiency of the locomotive may be obtained from the formula

$$A = 9.81 \frac{G\,(f+s)}{n} \times v,$$

in which

A = number of watts per second of the work done by the dynamo,

G = weight of train in tons,

v = speed per second in metres,

s = grade in per cent.,

f = the traction·co-efficient, that is, the effort which is necessary in order to move it on a level track.

If the weight $G = 1$ t., and the speed per second $v = 1$ m., n is the efficiency, that is the quotient of the useful work done at the circumference of the wheels and the total electrical energy absorbed by the locomotive. In the above formula f and n are unknown; $A = 16,700$ watts; $G = 20.84$ t.; s (mean value) $= 5.7$ per cent. ; in 8 minutes 30 seconds 2,200 m. were made. From this $t\ v = 4.3$ m. If we put these values in the formula, then we find as an

approximation $\left(\dfrac{f + 5.7}{n}\right) = 19$.

In order to find n we must accept for f an approximate value. If we take $f = 7.5$, which is about the mean value between that of 2.5 found at the last Paris exposition with the Décauville narrow gauge road, and the value $f = 12.2$, found by Tresca for level routes on the electric car line between Paris and Versailles, a road which is kept in good condition, then $n = 70$ per cent.

Engineer J. L. Hubel has, however, found by numerous and careful experiments on a line of horse cars in Hamburg a larger value, viz.: $f = 15$, but the conditions on the Sissach-Gelterkinden road resemble much more nearly the first two lines when "Vignol" rails are used. The measurements given above had to be made with the volt and ampère meters in use at the station, and their constants had to be accepted as correct.

Comparing these results with those obtained in this country Dr. Denzler says: "Even if the real efficiency should be a little smaller than 70 per cent., as obtained above, still the great progress made here cannot be denied in comparison with the results obtained with the best known American systems. If we take. for instance, as a basis, the results published by Dr. Louis Bell on the electric road in Lafayette, Ind., then we get $n = 47$ per cent., even if we take Huber's larger value for $f = 15$."

Another measurement on a grade of $s = 5.3$ gave $n = 49.5$ per cent.

Similar results can be calculated from Crosby's experiments with reference to the roads in Cleveland, Scranton and Richmond. In no instance does the value of n exceed 50 per cent. This means that in every case about half of the expended energy is lost in the form of heat and friction when high speed motors with double transmission are used. The value obtained shows plainly that the recently applied method of increasing the efficiency of electric roads by reducing the speed of the motor is the proper one. The ideal in these regards is a motor car or an electric locomotive in which the car axles and the armature shafts are the same, so that all loss in gears is avoided.

Dr. Denzler also points out the importance in electric railway operation of deciding whether it is desirable to use independent motor cars or an electric locomotive with trail cars, and under what circumstances each system should be employed. In favor of the locomotive it is pointed out that at first larger electric motors may be employed, which can be constructed for slower speed and of better design than if it is necessary to build the motors for the limited space available between the surface of the street and the floor of an ordinary car. In an electric locomotive, too, the motors are more easily accessible and can be better cared for. Dust and moisture proof appliances can be employed more conveniently, and nearly every part of the apparatus can be constructed more solidly than when smaller machines are employed. On small roads like the one described above, light trains are sometimes needed and at other times heavy ones, and it is especially desirable to be able to attach freight cars to the trains at different times. For necessities of this kind there are no difficulties in the way of constructing locomotives of any desired capacity, as is the case with independent motor cars, where the space to be occupied by the motor is limited. In this latter case it is not possible to go beyond a certain size of motor, and when larger trains are needed two or more motor cars must be used, in which case additional employés are needed, although the demand for more cars may be confined to Sundays and holidays. Interest on the capital invested, maintenance and attendance, will cost more than when a single locomotive of large capacity is employed. The disadvantages which are connected with the use of special locomotives are pointed out by Dr. Denzler to be as follows: The length of the train is greater than when independent motor cars are used, a disadvantage which would be especially felt in narrow business streets. The first cost of an electric street railway in a city where many cars are run at the same time will be greater if electric locomotives are employed than when independent motor cars are used, while repairs and maintenance will be rather less with locomotives. For these reasons it is impossible to select either of the systems as the best for all cases. The advantages and disadvantages of each system must be carefully considered in each particular case in connection with the local circumstances and demands.

On the electric road described above nine regular trains are run daily in each direction; for frieght distribution special trains are run. The time consumed in running from Sissach to Gelterkinden is 15 minutes, including stops. The stations on the line are connected with the power lines by telephone, the telephone line having a metallic circuit, the wires of which are run upon the same poles that carry the feed wire for the trolley line, and it is said that only a slight humming sound is perceptible on the telephone line when the train is in motion along the line.

Siemens & Halske Roads.—In place of the usual trolley, which they claim jumps out from under the wire too frequently, the firm of Siemens & Halske (Berlin) use a horizontal wire four or five feet long, running crosswise to the car as seen in the adjoining illustration; this is pressed up against the overhead line, against which it slides; the contact wire is attached to a single bar, making a T-shaped figure, the upper part having a bearing against the line and the lower part being hinged and overweighted at its lower end. Besides being a sliding instead of a rolling contact, and avoiding the rupture of the contact by a jumping out of the usual trolley wheel, they claim as additional advantages that the suspension of the line wires thereby becomes much simpler at curves and switches, in which they are doubtless right. On straight stretches the line wire must be suspended slightly zigzag, so as not to cut grooves in the contact piece. An objec-

FIG. 23.—SIEMENS & HALSKE CONTACT DEVICE.

tion to it is that it wears out the line wire and that it makes a disagreeable hissing noise, which is transmitted along the wire far enough to be distinctly audible almost one square each side from the car. The line wire is five metres (16 feet) above the track, and is suspended between the tops of wrought iron tubular poles made by the Mannesmann process. The line wire is a six millimetre (full No. 3 B. & S.) iron wire, continuous throughout; it is connected at every 50 and 60 metres to a feeder.

Thomson-Houston Freight Locomotive. — The freight locomotive, shown in the accompanying illustration, weighs 21 tons, and is said to be the largest electric locomotive ever built in this country; its general appearance is that of a small steam locomotive without the boiler. The massive iron frame consists of two heavy pieces of cast iron bolted together at the ends. The end plates are provided with heavy spring drawbars such as are used on the heaviest freight cars, while a cow catcher is cast solid with the end of the locomotive. One motor, having a maximum capacity of 125 horsepower, supplies the motive power. This motor is geared to the forward axle by a train of powerful cast steel gears. Parallel rods 1¼ inches by 4¼ inches connect the two axles. The distance between the centres of axles is six feet four inches, while the diameter of the drive wheels is 42 inches. The motor is made with a special waterproof finish throughout, designed for the purpose. The armature makes 1,000 revolutions per minute, when the car is running at a speed of five miles per hour. This speed is

FIG. 24.—THOMSON-HOUSTON FREIGHT LOCOMOTIVE.

reduced by the gears from 25 revolutions at the armature shaft to one at the axle. All the gears are of steel, and entirely enclosed in tight iron cases carrying heavy grease. The largest of these, that attached to the axle, has a width of eight inches. A very powerful brake, consisting of a cast-iron drum keyed solidly to the intermediate shaft of the motor, is provided. This is covered with a wooden lagging, and the whole embraced by two steel bands which are tightened by powerful cams operated by a lever in the cab. This brake is sufficient to stop a train of eight cars when going five miles an hour down a two per cent. grade. Sand boxes deliver the sand at the rails directly under the wheels on both sides of the locomotive, thus securing maximum traction effect. In the cab, or operating stand, is placed the controlling mechanism, consisting of a rheostat, reversing switch, brake, and sand box levers. The current is taken from a trolley wire by means of what is known as the "Universal" trolley, the peculiar feature of which is that when the locomotive is reversed it is not necessary to reverse the trolley. This locomotive is to be used for hauling freight cars from the factory to the main line of the railroad, a distance of two miles.

Some data on the locomotive are given below:

Voltage required for the motor............................ 500 volts.
Horse power at the drawbar................................ 100.
Speed on level track when developing this power......... 5 miles per hour.
Wheel base... 6 feet 4 inches.
Diameter of wheels....................................... 42 inches.
Speed reduction between armature and axle................ 1 to 25.
Gauge.. 4 feet 8½ inches.
Wheel base... 6 feet 4 inches.
Measured height above rail platform...................... 4 feet 4 inches.
Greatest length of locomotive (at cowcatcher)............ 15 feet 7½ inches.
Greatest length of platform.............................. 7 feet ¾ inch.
Weight of locomotive less trolley pole................... 42,525 pounds.
Approximate weight of motor.............................. 5,400 pounds.

In a number of trials, the first test was made by coupling the locomotive to two cars heavily loaded with iron. These were shifted from one track to another, both tracks being on a 2 per cent. grade, and also on a curve of about 150 feet radius. The total weight of the cars was 54½ tons. The second test was made by coupling on two more cars, making four in all, and shifting them as before. The total weight of the load was 96 tons. The third and final test was made with six loaded cars handled in the same manner, and with apparently as little effort as was required to shift two cars.

Another freight locomotive built by the same parties was illustrated in *The Electrical World*, June 20, 1891, page 462. It differs from this one in that its general appearance is that of an ordinary platform freight car, measuring 8x18 feet, and can therefore be used as a car besides being a locomotive. It is capable of hauling a load of 60,000 pounds at a rate of five miles an hour. The single motor is of 30 horse-power.

CHAPTER VI.

CONDUIT AND SURFACE CONDUCTOR SYSTEMS.

Under this heading are included those systems in which the contact wires are placed in conduits, open or closed, or those in which the two rails are used, as the two conductors. While much is hoped for from this system, with one exception little has been accomplished in actual practice. The road running in Budapest, described below, is the only large plant of this kind, and it appears to be running very successfully. The introduction of this system into America appears, from reports, to be in the near future. While the Americans have been losing time in endeavoring to reduce the prejudice against overhead wires in large cities, this European firm complied with the requirements of municipal authorities, and has constructed a successful conduit system. The features of conduit and rail systems are best shown from the following extracts taken from the published expressions of writers.

In a very interesting article on electric railroads, Mr. F. L. Pope states: "The original invention of Field contemplated the supply of electricity to the traveling car from conductors inclosed in a conduit beneath the pavement. He, as well as many other inventors, appreciating the force of that prejudice which undoubtedly exists—among newspaper editors--against any avoidable multiplication of overhead electric wires in the streets, realized that the great prize to be sought for above all others was the invention of a system of electrical supply which should dispense altogether with the overhead line. Hosts of inventors have diligently wrought for ten years upon this most difficult problem. Hundreds of patents have been taken out and more than a million dollars have been disbursed in paying for tuition in the costly school of experience. More than once, and in more than one direction, success has at times seemed almost certain; yet the truth compels me to say that from the hard practical standpoint of dollars and cents, by which every invention must first or last be tried, the net outcome of all this vast expenditure of labor, time, and money has, up to the present moment, been almost insignificant. The reward which awaits the fortunate person who succeeds in completely solving this problem may well be regarded as a potentiality of wealth beyond the dreams of avarice. The problem of the underground conduit does not at first sight appear to be a very difficult one. It renders necessary, in the first place, a construction which will effectually resist the action of forces tending to disturb the condition of the wires, and with the heavy traffic on the streets this involves a very strong structure. It is absolutely necessary that the conductors shall remain insulated from each other and from the ground, under all conditions of weather. The exigencies of heavy rains and snows necessitate a construction which shall permit of a thorough insulation of the conductors and a drainage of the entire system. There are other minor points which require to be taken into consideration. Without going into details, it is sufficient to say that the conduit system has been tried on an extensive scale in Denver, Cleveland, and Boston, and to a lesser extent in several other places, but in every case the continual interference and interruptions consequent upon its use have exhausted the patience of the traveling public and compelled its abandonment. I do not wish for one moment to convey the impression that the problem is an insoluble one. I may, perhaps, say that I think great progress has been made in this direction within the past year, and I cannot but feel that the disappearance of the overhead system from the streets of all our larger cities and towns is only a question of time, although the objections which have been most strenuously made to its introduction are for the most part rather of a sentimental than of a practical character."

The following opinions were given by Mr. Mansfield; we would add, however, that he does not appear to be aware of the fact that there is such a system running with apparent success in the city of Budapest: "There have been in this country at least four practical experiments with conduit roads, several hundred thousand dollars have been expended in testing it, and thousands additional in perfecting it, particular attention being given to the protection and insulation of the bare conductor. In spite, however, of all this refinement and study, practically nothing has been accomplished; and I have no hesitation in asserting that the continuous live conductor in an open slotted conduit is to-day a failure, and that it cannot be made a success throughout our cities of to-day, its fatal weakness being our inability to prevent the conduit from becoming filled with water, mud, etc. The time may come when our sewerage systems will be perfect enough to enable us to overcome this fatal weakness. To-day, however, they are not, and even an optimistic view puts this time a long way distant.

"The inventions covered by the surface method are somewhat similar to those employed in the conduit system, only, in place of being in a conduit, they are placed upon the surface of the street. By far, however, the larger number of arrangements are based upon what might be called the "interval" or "point" system. Primarily this arrangement con-

sists of an underground insulated conductor connected by means of taps to contact points on the
surface of the ground, held in place by means of
iron boxes, and insulated therefrom by means of
rubber, wood, fibre, or other similar substances.
Upon the car is swung a long contact plate extending practically from one end to the other. This
plate is carried close to the ground, and is arranged
to touch the points as it passes along. Many automatic arrangements have been devised to cut these
contact points into circuit by the car as it passes
along One inventor had electro-magnets on his car,
which imparted their magnetism to the iron contact
pins or points. These becoming magnetized would
lift a little armature inside the iron casing, which
in turn moved a switch and cut the iron contact
points into circuit. Another inventor arranged a
slot between his tracks, in which the plow passed.
As this plow passed the contact points, it would tip
a lever or some other automatic device, which would
throw a switch and put the iron contact points into
circuit. All these arrangements have the same fatal
weakness—a liability to ground or short circuits. I
do not consider any of them practically possible.
There is one more system, which practically com-
bines all these methods, and which obviates many
of the objections. The inventor has a slot between
his rails and boxes, with contact device within,
placed at proper intervals. Upon his car is a plow
which passes along through the slot, and also a long
contact plate or arrangement extending its entire
length. The operation of the invention is as follows: The plow as it passes through the slot strikes

a lever placed in connection with each box, which lifts for a distance of six or eight inches above the ground a piston carrying the contact piece proper. This is made in the shape of a right angle hook placed within a vertically moving piston, and thoroughly insulated from it. As this is raised up, the long contact plate under the car passes beneath the hook and holds it up. The current is taken into the car as the plate slides under the live hook. Naturally, as the car passes along, the hook slides off from the end of the contact bar and drops back into place. To protect this hook, or contact piece proper, it is covered with an extension of the cylinder, so that, as far as the street is concerned, the surface is perfectly smooth, and one sees nothing but a small round cover in the centre of each of these boxes. The contact hook is alive only when it is resting on the contact plate of the car. Certainly, in so far as getting rid of all the troubles due to the street being covered with water, this is successful. The fatal weakness whereby the contact points remain permanently on the street surface is obviated here by the contact point being practically lifted six or eight inches above the surface. In regard to the permanence and reliability of this system, I can say nothing, as no trials have been made.

"I have no hesitancy in asserting that this class will never prove a success, nor can it be made to work successfully throughout any of our cities to-day. The reason for this is obvious. It is immaterial whether the automatic devices prevent the sections or contact points from being alive all of the time or not; if the conduit is filled with water or mud, and the points are made alive just as the car passes, there is bound to be a momentary grounding from these points. In other words, a point made alive in water with the other side of the circuit grounded, is just about as dangerous and bad as if it remained alive. It is true that the grounding may not be as severe, but still it will occur, and with a large number of cars throughout the city moving at the same time, causing therefore a large number of points to be alive, if any number of points were grounded through water it would cause a tremendous loss upon the central power station, and in all probability a complete grounding or short circuiting of the entire system. Numerous ingenious schemes have been devised to overcome this fatal weakness. None, however, have ever been put to trial. I have seen many, yet to-day there is no hope in this direction.

"Several of these devices are very interesting and worth mentioning. One inventor placed his bare conductor in a rubber tube made something like a hose pipe. Arranged upon the upper side of this tube are the contact points, which might be likened to rivets punched or tapped through the tubing. The contact wheel or device rolls along the top of this tube, and is sufficiently heavy as it proceeds to press the tube together so that the contact points or rivets come in contact with the conductor inside the tube. By this means the current is transmitted to the motors. Obviously, as the car proceeds, the tube behind the contact wheel springs back into its original position.

"Another arrangement was to have inside the

conduit a large elastic tube with a bare conductor inside, this tube to be slit along its upper side. Attached to the car is a plow with a contact device at its lower end inside the tube and in contact with the conductor. The plow moves along with the car and slides through the slit. With this arrangement it is apparent that if the conduit were filled with water it would run in just before and behind the plow. Should this occur, the old fatal trouble would be bound to follow. Furthermore, it would be exceedingly difficult to arrange the tube sufficiently rigid in position to withstand the action of the plow, and the sides of the slit sufficiently tough to withstand the friction. The question of turnouts would also be serious and troublesome.

"Another inventor arranged on each side of his conduit slot rubber flaps shod with steel and held together by means of springs. Attached to the car is the plow as usual. As the plow passes along in the conduit it presses back the flaps. The objection to this is, of course, the liability of the water getting into the slit before and behind the plow. To obviate this, the inventor ingeniously suggested that the whole conduit be kept under an air pressure, which air pressure would be sufficient to blow the water out at any of these open places where it tended to run in. I do not consider that any of these plans can ever be successful.

"Summing up the general results of the underground and surface methods, it certainly looks as if we could not expect very much from them in the immediate future. Much perseverance and money must be expended before as practical and certain a

method as our present overhead one is · attained.
We surely are all anxiously and hopefully waiting
for it."

Mr. F. H. Monks says: "The conduit system has
been thoroughly tested, and has been shown to possess no commercial value for the propulsion of street
cars to date."

Descriptive.

Besides the few systems incidentally described in
the above extracts, the following may be of interest:

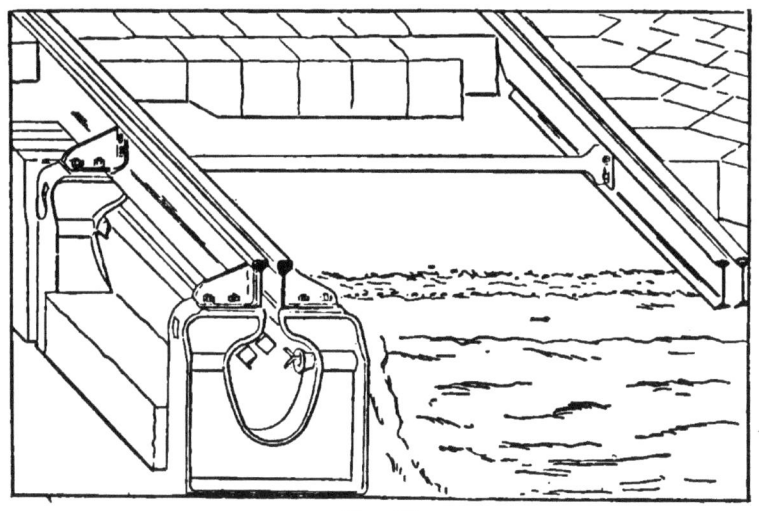

FIG. 25.

Siemens & Halske Conduit System; Budapest
Railway.—Undoubtedly the most interesting European street railway system to the Americans, who
can learn little from abroad regarding the overhead
plan, is the underground conduit system of Siemens
& Halske, as introduced by them in the city of
Budapest, the capital of Hungary.

The conduit, as seen in Fig. 25, is placed under one rail. It consists of castings having flanges of 18 centimetres (7 inches) placed every 1.2 metres (about 4 feet), the space between being a conduit of concrete. The oval shaped conduit has a width, clear, of 28 centimetres (11 inches) and a height of 33 centimetres (13 inches).

The slot consists of two beam rails having no inside lower flange, and fastened to the conduit frames by wrought-iron angle pieces. The width of the slot is 33 millimetres ($1\frac{5}{16}$ inches). The total depth of the foundation below the rail top is 70 centimetres ($27\frac{1}{2}$ inches). The conductors, both positive and negative, are made of angle irons, secured, as seen in the figure, by means of insulators fastened to the castings. They are sufficiently high above the floor of the conduit to be protected from the water which may collect in the conduits. They are, furthermore, under the top of the oval, so that they cannot be touched from the outside. It should be noticed that there is no earth return used in this system, as both leads are insulated. The water which runs into the conduits is collected at the lowest points and passes through settling boxes to the sewers. The second track may be of any desirable form, even only a flat rail. An objection to having the conduit under one rail instead of in the middle arises in cases where, by the nature of the course of the track and the curves, the cars become reversed in their positions on the track; such cases can probably be avoided by proper laying out of the road, and in the worst case by a second conduit under the other rail for parts where it cannot be avoided. On a one-track line the cars must have their front and back platform alike, as they cannot be turned around.

The car truck and the motor are the same as those used in connection with their overhead system; they will be found illustrated under the heading of motors and trucks. Only one axle is driven, the other is flexibly connected to the third point of support of the motor frame. The starting, stopping, and reversing of the motor is done as usual by a crank at either end of the car, operating the same switches. Controlling resistances are placed under the body of the car.

An illustration of the generating station, too large for reproduction here, may be found in *The Electrical World*, October 24, 1891, page 304. This station contains three compound condensing steam engines of 100 horse-power each; in another part there are two more of 200 horse-power each, driving direct coupled dynamos. The system is run with 300 volts, and all the dynamos lead to common bus wires. The leads are made of lead covered cables armored with iron bands and laid directly in the earth; these lead to junction boxes from which others run out along the road and are connected at intervals by means of short branch cables to the iron leads in the conduits.

The greatest up grades are 1.5 per cent. and 1.6 per cent., and the smallest curves have a radius of 50 metres (168 feet), 25 metres (84 feet), and 45 metres (148 feet); the latter is on a 1.6 per cent. up grade. The allowable speeds are as follows: 15 kilometres (9.3 miles) in the city, 18 to 20 kilometres (11 to 12½ miles) in the ontlying portions, 10 kilometres

(6.2 miles) in the densest parts of the city, and six kilometres (3.7 miles) over street crossings. There are at present 20 kilometres (12.4 miles) of road in use, and 50 cars. Each car travels daily 120 to 130 and even 150 kilometres (75-93 miles) in a 16-hour service.

The first part of the road, 2.5 kilometres (1.55 miles) long, was opened July 30, 1889, and has, therefore, been running over two years. The system has been so satisfactory that it is being extended quite rapidly. It was built at the expense of the firm of Siemens & Halske, and may, therefore, be said to be a practical experimental line of that company. They state that some Americans have been negotiating with them to introduce this system on a large scale in one of the principal American cities, but the name of this city was not given.

The following figures from the official reports of 1890 may be of interest, as also the deductions which can be made from them. There were running during that year three lines. The total number of passengers carried was 4,459,234; the total car kilometres were 758,838.1 (about 470,000 car miles); the total income from fares was 275,742.97 florins (about $113,000). From this it follows that the average number of passengers per car kilometre was 5.88 (9.48 per car mile); the average income was 36.27 kreutzer (about 14.8 cents) per car kilometre (about 24 cents per car mile); the income per passenger, that is, the cost of the ride to the passenger. was 6.18 kreutzer (2.53 cents). There were 6 kilometres of line in operation at the beginning of the year and

9.1 at the end. In a public address, one of the engineers of the company stated that the latest figures (July and August, 1891) of cost of running were only 37 per cent. of the income, which certainly is remarkably low. The city tax, which is very high in Budapest, is not included in this.

Comparing this with the corresponding official figures of the horse-car lines of the same city, we find that the total number of passengers carried was 18,107,543; the income from fares was 1,485,180 florins (about $609,000); the income per passenger, that is, the cost of the ride to the passenger, was therefore 8.20 kreutzer (about 3.36 cents) as compared with 6.18 kreutzer for the electrical roads. The fares of the electrical roads are therefore cheaper, being 24.6 per cent. less than the horse lines, or the latter are 32.7 per cent. more expensive to the passenger than the electric roads. The frequency of the cars per kilometre of line was larger at the beginning of the year than on the horse-car lines, and at the end was double the other. The income of the electrical road per kilometre of track was more than 1.5 times that of the horse cars in the latter half of the year. The latter figure, however, depends on the traffic in the particular districts through which the roads pass and is not necessarily a criterion in favor of electric roads; the income on the electrical roads, however, increased from 1,153 florins per kilometre of road in January to 3,621 in December, while that for the horse car roads remained about the same in December as it was in January, namely, 2,267 against 2,124. The length of the horse-car roads was 45.6

kilometres in January, and 45.8 in December.
Unfortunately the report does not give the number
of car miles of the horse-car lines, so that no comparison can be made with that figure. It should be
noticed, however, that the length of the horse-car
lines was five times as great in December as that of
the electric lines, and it is therefore likely that the
lengths of the rides were greater than in the short
electrical roads. This is all the more likely, from
the fact that the number of car miles of the horse
cars was omitted in the report.

Love Conduit.—This system consists briefly of a
small conduit 15 inches deep, placed between the
rails, as in the cable system, and having a slot. The
slot rails forming the top of the conduit are
arranged so as to be detachable, thus permitting
the conduit to be readily inspected and cleaned.
Two bare copper wires are secured on insulators
near this slot, but are protected by a deep flange
which protects them from any interference through
the slot. Two illustrations of this conduit will be
found on page 9 of *The Electrical World* of July
4, 1891.

Gordon System.—Like in the Lineff system, this
consists of a closed conduit, the top of which is composed of metallic sections which are in circuit only
when the car passes over them. It was described in
the London *Electrical Engineer* as follows:

"The conduit carrying the current is a very small
one, some two inches or three inches of ground being
quite sufficient to contain it. The supply rail is
laid midway between the two line rails, and consists of flat iron laid in concrete in lengths of about

eight feet, or one-third the length of the car. The system arranges for charging these sections by the full current of, say, 400 volts, as the car progresses, so that no section is charged except those under the car. This is done by a system of connections laid in a gas pipe with T-pieces connecting to each length and leading back to a commutating box, which is the feature of the system; the conductor sections are charged by a return insulated circuit actuated by magnets placed, not in the car, but in a box under the pavement. These boxes are placed every 100 yards under the curb, and contain strong long-pull magnets, one for each section. As the car progresses, a shunt current comes back from section No. 1 to magnet No. 2, which rises and puts section No. 2 into connection, cutting off No. 1, and so on, as the car moves—the main, of course, running the entire length of the road. The pull of the magnet is about seven pounds, and evidently is more than ample to do its work. The advantages of the system are, the use of the closed conduit, no slot or wires that can be touched being needed, the small cost of conduit and corresponding small cost of alteration of track. Another advantage is that by the addition of one contact the absolute metallic return for the current can be made by way of the section behind, and thus the difficulty with the telephone companies can be avoided—a matter impossible of arrangement with most systems without costly additions to the mains. The cost of alteration of track and laying the electric conduit is estimated (the figures being checked by Messrs. Merryweathers' engineer) as £2,000 per mile, of which £25 each is the cost of commutator boxes. The cost of a new electric car would be £400. He estimates the actual cost of traction at 2½d. per car mile."

Pollack System.—In this system there is a middle contact rail in short, insulated sections; the conductor carrying the current lies directly beneath this rail, and by means of flexible contacts momentary connections are made from it to that particular section of the contact rail over which the car is at that moment. These temporary connections are made by the aid of a magnet on the bottom of the car, which, in passing over, attracts iron blocks under the rails, making contact with that section. It will be noticed that this is similar to the Lineff system. Pollack claims to have invented it in 1886, though not in just this form, which he devised a year later. The iron contact rails are double, in order that the magnetic circuit shall be open at the bottom, thereby developing a greater magnetic force there than if a single band of iron was used, through which it would be more difficult to attract a mass of iron on the other side of it. Permanent magnets are to be used, having in addition a winding through which the current may be passed to reinforce them. The independent spaces inclosing the movable blocks, being entirely inclosed by asphalt, may be kept dry. As there are two contacts to each section, it is stated that there is no sparking at these contacts, as the current is never interrupted there. In addition, he proposes to provide the car with a device (presumably some accumulators) to drive it back on to the rails in case of derailing. This system is not yet in use anywhere. If these contacts can really be kept dry and free fiom sparks, this system looks like a very simple solution of the problem of an underground supply for the current.

Schuckert System.—A modification of this was proposed by Schuckert & Co., of Nuremberg. It differed in the nature of the contacts, which were made of iron filings contained in little chambers under the rail. When attracted by the magnet on the car they bridged over from the main to the contact rail, making the necessary connection. The conductor sections are laid along in the roadway upon solid wood planks, imbedded in the ground, under which is the main strip conductor. Each wooden plank is pierced downward at intervals along its length by taper holes of $1\frac{1}{2}$ inches to 2 inches diameter, which have the smaller diameter at the top. Before laying the iron conductor rails over the planks, these holes are each partly filled with a handful of iron filings, which drop down upon the main conductor. As the tram passes, its magnet energizes the roadway beneath it, and the iron filings rise in a heap and make contact with the surface conductor, falling again as the car passes. The taper form of the hole prevents their continued contact after the car passes, and the number of holes and the mass of filings is said to obviate all difficulty with reference to contact or sparking. It is claimed that in this case if there should be sparking it would do no harm, as the iron filings would always present many new contacts. It is not in operation at present. It was noticed in an experimental line that, even when the rest of the street appeared dry, there was considerable moisture in the narrow section of asphalt which separated the two ends of the sections, quite sufficient to conduct considerable current from one to the next, thereby doing just that which this system was intended to avoid. The iron contact rail was in this case a single flat band of iron, and not double as in Pollack's system. It is a question whether the iron filings will behave as they are intended to behave. It is ingenious, but can hardly be called a good mechanical construction.

Edison's Proposed System.—During the month of October, Mr. Edison made some statements to the reporters of a daily paper, regarding a new system which he is about to introduce. His published description is so meager and unsatisfactory, and the language used is such, that only a very rough idea can be obtained from it. We give below some points gathered from these short reports as well as from other sources.

He claims that he will be able to obtain a speed of 100 miles an hour with great ease on a 100-pound rail, on a rock-ballasted track. He claims that it will be the cheapest system known, and that the plant will not cost half as much as the cable on a cable line. He claims to be able to lay a mile of track in a single night. He furthermore expects to get one horse-power from one to two pounds of cheap coal, while a steam locomotive requires six pounds of expensive coal.

He intends to begin with the Chicago and Milwaukee road, at the time of the World's Fair, on which he intends to run a train of two cars every 20 minutes. He claims that there are special engines with a horse-power of 10,000 and 12,000 each, which would run the whole Pennsylvania Railroad system, between New York and Philadelphia, including freight, local and express trains, at a reduced expense, with much less depreciation of rail stock and roadbed. There are said to be 400 or 500 engines on the Pennsylvania Railroad at one time. He furthermore stated that he will soon have a track for demonstration ready in the vicinity of New York.

The *Edison Monthly Record* for December states: "Mr. Edison has devised the new system for roads of heavy traffic, in large cities where the expense of the original installation is warranted by the traffic, and where the trolley system will not be permitted. For instance, the new system would not be applicable, in a commercial sense, to long roads operating less than fifty cars simultaneously. It must therefore be understood that, outside of the large cities, the best system that can be advocated is the trolley."

The following particulars have been furnished by Mr. Edison: "The overhead system is entirely dispensed with. Cars, trucks, tracks, and roadbed, such as are now in use, are retained, certain changes being made in the joints and cross-ties. The power, furnished by 1,000-volt generators, is distributed to reducing apparatus placed in boiler-plate manholes at intervals varying in accordance with the number of cars required to be operated. At these various reducing points the current is transformed from 1,000 volts to a pressure of 20 volts, and put in direct communication with the tracks. This limit of 20 volts is fixed in order to prevent horses from being affected by the current. The economy of current is about the same as with the present system of trolley.

"The car motors being wound with uninsulated copper wires, and the pressure of current being so low, there is entire freedom from burning out of armatures, as water can be poured upon the armatures without any ill effect. The problem of producing a perfect rail joint, and the picking up of a heavy current from a mud-covered rail, has been solved in a practical manner. The experimental road at Mr. Edison's laboratory is a quarter of a mile long, with a six per cent. grade and very short curves. It is operated successfully when the rails are entirely buried in mud or in dry sand.

"Mr. Edison is arranging to have the system placed in practical operation on a heavy traffic road in some large city (probably New York) to demonstrate its practicability. This will be done during the coming year."

The effect which the announcement of this system has had on the electric railroad industry was a very harmful one, as it made many railway companies hesitate about adopting the present systems. Unless, therefore, Mr. Edison can substantiate what he claims, he has done great injury to the industry, as well as to his own reputation.

Commenting on this system, Professor Elihu Thomson says: "Of course every electrician knows as well as I that the conditions of practice in street railway work are not such as are likely to permit any such scheme to have a ghost of a chance to survive. Certainly such statements, coming from Mr. Edison, are calculated to retard the growth of the industry, particularly as the schemes which he puts forward as the solution of the problem of street railway work have not reached the beginning of a demonstration of their feasibility, a demonstration which is hardly likely ever to be made."

Professor Sydney H. Short says: "If the pressure is so low that it will do no harm to vehicles or animals coming in contact with the conductor, it will be impossible to keep contact with the rails through dust, mud, snow, etc.; it is now a very common thing for a car operating with 500 volts pressure to become stranded by being insulated on dry dust or on snow. The question of resistance comes in also, as a very important one. The resistance across four or five feet of wet earth from one rail to the other in a track several miles long must of necessity be very low, so low that I think it would practically short circuit the rails. It troubles me to think of the size of the conductor that would be necessary to carry the enormous volume of current needed in such a system. While I am willing and anxious to believe it possible to operate street railways in the manner described, my experience has been such as to make me incredulous."

Mr. O. T. Crosby states: "Plainly stated, the problem of high speed electrical work does not demand a genius. It demands good railroading, good engineering, and plenty of money. A current flowing through each rail, in case there be a heavy traffic, will be at a point near the secondary generating station (at the motor dynamos) anywhere from 1,000 to 3,500, or even a larger number of ampères. What this will do at the rail joint remains to be seen. It is also a subject of easy calculation, to determine how much copper must be added to the rail in order that the loss, at 20 volts even, from the short distances in question, shall not be considerable. Metal tie bars joining one rail to the other must be given up, or be used with non-metallic bushings. At switches, turnouts, crossings, etc., the problem of preventing metallic contact between rails on opposite sides of the track would be one which would startle almost any practical track builder who knows the difficulties of using anything but metal in track construction. Concerning the matter as a whole, that is, the transfer from high to low potential, and the use of the rails as conductors, it presents such an enormous burden in the shape of first cost of plant, together with such unusually large transmission losses from primary engine to the motor on a car, and requires such apparently impracticable changes in ordinary railroading methods, that in my opinion nothing save the magic of Mr. Edison's name could obtain for it any hearing from the interested public."

CHAPTER VII.

STORAGE BATTERY SYSTEMS.

From the published expressions of a number of writers it appears that the opinion is almost universal that the storage battery system is the "ideal" system; but at the same time it is also their almost universal opinion that it has not yet to-day been developed to a perfectly satisfactory system. The fact that such roads have been abandoned, that very few are in operation, and that still fewer are contemplated does not appear to be a favorable state of affairs. The objections, however, seem to be all traceable to the defects of the batteries and not to the system, notwithstanding the fact that such a system requires the carrying about of its source of power. The steam locomotives and the horse cars have to do the same. It will be seen from some of the plants described below that the weight of the battery, though large, is not a very large proportion, being in one case only about 23 per cent. of the total load. The fault appears to lie in the fact that the batteries cannot stand the strain of the heavy output required of them in starting the car, and to the fact that they deteriorate too rapidly. As soon, therefore, as the battery is perfected, storage battery traction will take a much more prominent place among the other systems.

Among the numerous opinions expressed during the past year on this system, the following extracts will give a very good summary of what is thought about it.

Mr. F. L. Pope, in a very interesting lecture, made the following remarks on this subject: "The storage battery system, both from the standpoint of the public and of the street-car manager, is an ideal solution of the problem of electrical transportation. It has many features which especially commend it for city work. Each car is an independent unit, and hence no accident can materially cripple a well organized system. The idea of being able to store a large quantity of electricity in a box under the seats of a car, to carry it around, and to draw from it a large amount of power, as required, is a most attractive one. Storage battery cars can be introduced in the existing systems, one car at a time. In cases of emergency they can be run over almost any route where a horse car can be . run, not being restricted to the route of an electric conductor, and under ordinary circumstances the car can be run just as rapidly and controlled just as effectively as it can be with the trolley system.

" All these advantages were recognized at an early day, and hardly had the storage battery been invented before it was applied to street-car propulsion. The first storage car was operated in Paris, in 1882. The system was put in practical service on one of the lines in Brussels, Belgium, and continued for two years. It was also introduced in 1886, on the Madison avenue line in New York city, and has been in operation there ever since. The results of the ten cars now running on the Madison avenue line appear to have demonstrated that the system may, with careful management, be made thoroughly reliable. Each car will carry a full load on a straight and level track 15 miles per hour, and will ascend grades not exceeding five per cent. and not more than 500 feet long at five miles per hour. A run of 40 miles can be made on a level track, which is considered a half day's work, with one charge of battery. The time required to replace the exhausted batteries with fresh ones is not more than two minutes. There are, however, two serious objections to this system, which have thus far been sufficient to prevent its adoption in other than exceptional cases. One of these is the great weight of the batteries, and the other is the lack of reserve. power in the emergencies which sometimes occur. In other words, two considerations enter into the question, one of economy and the other of efficiency. To take up the question of comparative efficiency first, a very little consideration of the matter will show that the progress of the art, so far as yet developed, practically limits the use of storage battery cars to lines having very light grades—not more than five per cent. at the outside. Very many of our eastern cities, for instance, Springfield, Providence, Worcester, Albany, Troy, and many others which I might mention, have very severe grades, from 7 to 10 per cent., and some of them half a mile long. To start a heavily loaded car in bad weather on such a grade as this means an expenditure for a few seconds of from 50 to 80 horse-power. Experience has shown that this is far too heavy a tax upon the powers of a storage battery. Even on comparatively level lines, in our northern climates, the occurrence of a heavy storm of snow or sleet necessitates the expenditure of an abnormal amount of power, and it is on just such occasions as these when everybody wants to ride, so that the cars are loaded to their utmost capacity, and great dissatisfaction prevails if there is any delay or any lack of sufficient accommodation. It is at such times as this that the advantages of a distribution from a central station are most apparent. The whole number of cars, or extra ones, if necessary, can be run, and any amount of power required to force them through obstructions and to keep them running on schedule time can be supplied from the power station. It is merely a question of burning more coal.

"Considering the economical aspect of the question, first, as to the original cost, it may be said, roughly, the difference between the storage and overhead line system is not very great. The whole outfit, exclusive of roadbed, track and building, for either system, will aggregate perhaps $10,000 per car. Now as to the cost of maintenance and operation:

"A 16-foot car with two motors, such as are most commonly used, will weigh 8,000 pounds, of which 4,000 is electric apparatus. A storage battery car of the same size and capacity weighs 14,000 pounds, of which 3,800 is battery. A full load may be estimated at 30 passengers, weighing about 4,000 pounds. It will be seen that the storage battery car weighs nearly as much as two ordinary electric cars. It seems at present impracticable to operate a car with less than 3,800 pounds weight of battery, and unless great improvements are made there is little reason to hope that this weight can be materially reduced.

"What is termed the commercial efficiency of an electric railway is the ratio between the power generated by the steam engine and that exerted upon the car wheels. We may call the weight of the car body and truck, plus the full load of passengers, the useful load. The best authorities seem to concur in the opinion that it requires, roughly speaking, nearly twice as much power per unit of useful work done, with the storage battery system, that it does with the direct distribution system. The dead weight of the battery is nearly equal to that of a paying load of 30 passengers, and this load must be transported whether there are any passengers or not. The increased weight adds to the cost of renewal and repairs, both of the track and of the cars.

" The cost of horse power for drawing cars has been found by long experience to vary from 10 to 11 cents per car per mile. The cost of power for the ten storage battery cars on the Madison avenue line in New York is said to figure out 10.6 cents per car mile, practically the same as horse power. A careful computation of the average cost of power by the direct distribution system, on a considerable number of different lines in this country, shows that the cost for coal, attendance, real estate, oil and waste, repairs of machinery and line, foots up 5.09 cents per car per mile. I am now speaking solely of the cost of power; the greater part of the expenses of any street railway company remain substantially unchanged, whatever kind of power is used.

"The total number of electric cars now running in this country is probably between 2,500 and 3,000. Of the whole number, I presume not more than 30 or 40 are operated by storage batteries. The fact that the overhead system, though introduced at a later period than either the storage system or the underground distribution system, has so far surpassed them both, goes far to show that as yet it is the only system which has been able to meet the various exacting requirements of our street railway service."

In a report on the various systems, Mr. Knight Neftel states the following regarding storage battery systems: "Owing to litigation and a variety of causes, the storage battery system has made but little material progress in a commercial way during the past year. So far, primary batteries have been applied only to the operation of the smallest stationary motors. Their application in the near future to traction may, I think, be entirely disregarded. Were it not a purely technical matter, it might easily be demonstrated, with our knowledge of electro-chemistry, that such an arrangement as an electric primary battery driving a car is an impossibility. In view of the claims of certain inventors, I regret to be obliged to make so absolute a statement, but the results thus far have produced nothing of value.

"The application of secondary or storage batteries to electrical traction has been accomplished in a number of cities with a varying amount of success. Roads equipped by batteries have now been sufficiently long in operation to allow us to draw some conclusions as to the practical results obtained and what is possible in the near future. The advantages which have been demonstrated on Madison avenue, in New York; Dubuque, Ia.; Washington, D. C., and elsewhere, may be summarized as follows:

"First, the independent feature of the system. The cars being independent of each other, are free from drawbacks of broken trolley wires , temporary stoppages at the power station, the grounding of one motor affecting other motors, and sudden and severe strains upon the machinery at the power station, such as frequently occur in direct systems;

the absence of all street structures and repairs to the same, and the loss by grounds and leakages are also very considerable advantages both as to economy and operation.

"Second, the comparatively small space required for the power station. Each car being provided with two or more sets of batteries, the same can be charged at a uniform rate without undue strain on the machinery of the power station, and as it can be done more rapidly than the discharge required for the operation of the motors, a less amount of general machinery is necessary for a given amount of work.

"Another and important advantage of the system is the low pressure of the current used to supply the motors and the consequent increased durability of the motor, and practically absolute safety to life from electrical shock. It has been demonstrated also that the cars can be easily handled in the street, run at any desired speed, and reversed with far more safety to the armature of the motor than in the direct system. The increased weight requires simply more brake leverage.

"The modern battery, improved in many of its details during the last year, is still an unknown quantity as to durability. There is the same doubt concerning this as there was at the time incandescent lamps were first introduced. At that time some phenomenal records were made by lamps grouped with other lamps. Similarly, some plates appear to be almost indestructible, while others, made practically in the same manner, deteriorate within a very short time. It is consequently very difficult as yet to exactly and fairly place a limit on the life of the positive plates. Speaking simply from observation of a large number of plates of various kinds, I am inclined to put the limit at about eight months, though it is claimed by some of the more prominent manufacturers—and undoubtedly it is true in special cases—that entire elements have lasted ten months, and even longer. It must be remembered, however, that the jolting and handling to which these batteries are subjected in traction work increases the tendency to disintegrate, buckle, and short circuit; and that the record for durability for this application can never be the same as for stationary work.

"A serious inconvenience to the use of batteries in traction work is the necessary presence of the liquid in the jars. This causes the whole equipment to be somewhat cumbersome, and unless arranged with great care, and with a variety of devices lately designed, a source of considerable annoyance. The connections between the plates, which formerly gave so much trouble by breaking off, have been perfected so as to prevent this difficulty, and the shape of the jars has been designed to prevent the spilling of the acid while the car is running. The car seats are now practically hermetically sealed, so that the escaping gases are not offensive to the passengers.

"The handling of the batteries is an exceedingly important consideration. Many devices have been invented to render this easy and cheap. I have witnessed the changing of batteries in a car, one set being taken out and a charged set replaced by four men in the short space of three minutes. This is accomplished by electrical elevators, which move the batteries opposite the car, and upon the platforms of which the discharged elements are again charged.

"In the past four years a great change has come over everything connected with the storage battery as regards the plates, and the shape has changed completely, so that the storage battery of to-day, as manufactured by the leading companies, is an entirely new thing. I remember five years ago receiving a letter from the Thomson-Houston company, which I have preserved as a curiosity, asking me whether I thought electric traction would ever be successful, and whether it would pay an electric light company to manufacture motors for electric traction. In the light of the developments made in the past in the general field of electric traction, we should not cast discredit on the storage battery, or throw it aside as a failure.

"I think it is a mistake to expect batteries to make a long run. If we have a perfected storage battery system, it will not be a system in which our cars will run 100 to 200 miles; but it will be a system in which each car will make one trip, then newly charged batteries will be put in the car. The batteries can be changed easily, and there is no earthly object in dragging them along for 30, 40, or 50 miles; it is taking along an unnecessary weight in lead.

"The general conclusions which the year's experience and progress have afforded us an opportunity to make may be summarized as follows: Storage battery cars are as yet applicable only to those roads which are practically level, where the direct system cannot be used, and where cable traction cannot be used; and applicable to these roads only at about the same cost as horse traction. I feel justified in making this statement in view of the guarantees which some of the more prominent manufacturers of batteries are willing to enter into, and which practically insure the customer against loss due to deterioration of plates, leaving the question of the responsibility of the company the only one for him to look into."

Mr. E. A. Scott states: "I have seen the changing of the plates done in the Dubuque road in about 30 seconds, so that the question whether a storage battery car will run 100 miles without recharging, or 50 miles, does not become a matter of importance. There is no more necessity to run a battery 100 miles than there is of making a pair of horses pull a car that distance. You can work a pair of horses until they drop in their tracks, and you can do the same with a battery until it becomes exhausted, but there is no necessity in either case. We calculate to run about 20 or 25 miles before we make a change of batteries, and if they are run in that way in a road where grades do not exist beyond five or six per cent., there is no trouble. The question of how expensive the system is going to be is controlled by the question of how often you are going to renew the positive plates. We believe from what we know that the cost of running is considerably less than horses, although we have no figures to show. We believe that in about one year it can be run in competition, so far as economy is concerned, with any overhead system.

"When a car is run about 20 miles at the ordinary street-car speed, it will take about two and one-half hours to charge the batteries for that 20 miles or more; but practice has shown us that it is very foolish to draw out the entire charge of the battery. The question of how long the battery will run before it will absolutely break down ought not to be considered for a moment, because it is very bad practice—that is the thing which destroys the cells. If the cells are properly used in this respect, the batteries will last a long time, for many months. I know of one plant in Philadelphia in the lighting service that has run five years without change.

"As to the amount of power that is required to run a storage battery car, I took for one week the mileage of the cars on the Dubuque road, and also measured at the dynamo the amount of power put into the batteries. I was enabled to get from the mileage of the cars the amount of horse-power it took per car mile, and it was a little less than one and one-half per car mile. The road has grades of over six per cent; on a level road I have no doubt it can be done with less power. On one road it averaged for six months in a commercial way eight-tenths of a horse-power per car mile."

Mr. Crosby states: "No doubt the future has within it such possibilities as none of us, I think, should dare to measure to-day. I even go so far as to say that I believe in the future that it will be possible to propel a car by electricity without tying it by a string to the station. The trouble is we do not see how to do it at the present time. I think we can leave the matter for a time to those who are hard at work upon it."

Mr. F. H. Monks states: "Perhaps the ideal system of application is through the accumulator or storage battery. But can it be shown that any storage battery system has proved commercially successful after thorough and practical tests actually conducted for a period of 12 months upon a city road? Storage battery cars have been exhaustively tried in England for several years, and the latest reports from that country show that no degree of commercial success has been attained to date. The result in this country seems to be the same. Now and again we have been assured that a new invention in the storage battery field has been found, which will surely meet with commercial success in operation, but the report of actual tests and results in proof thereof is lacking. Personally, I have faith to believe it will come some day. Perhaps it is near at hand."

Mr. Montague states: "I have seen within three months a storage battery car that will take 85 passengers on a 16-foot car, and carry them up a grade of 10 per cent., and do it steadily and easily. It takes an hour and thirty minutes to charge the batteries. They do not allow the batteries, as a rule, to run more than 20 to 22 miles; but on one day three cars ran respectively 42, 46, and 48 miles, and when they got back to the power house the engineer had gone home, and they remained outside all night. In the morning there was sufficient power left in the batteries to take the cars into the power house. It takes about 30 seconds to change the batteries. These cars run on level ground at a speed faster than most city governments will permit, and in mounting grades they go quite fast. On one occasion when a car was mounting a grade with 30 people in it, I got out and walked alongside of the car, and not until it had reached the top could I keep up with it. The grade was about five per cent. There are six cars running now."

Mr. Barr states: "The Lehigh Avenue Passenger Railway, in Philadelphia, was built to operate storage battery cars. In May, 1890, the road started with six storage battery cars. In October, 1890, application was made to the Philadelphia Council for the use of the overhead system, claiming that unless they could use it they would have to abandon the road. The Council refused permission to the company to use the overhead wire, and on January 1, 1891, the storage battery cars were abandoned, and the road has since been operated by horse power."

Mr. Vhay states: "I represent a road which perhaps experimented with the storage battery system more than any other road in the country. The Woodward storage battery car was put on our road for about a year and a half. The system worked very well; the cars ran at the rate of about 12 or 14 miles an hour. The cost for coal was considerable, however—ten dollars a day—and it seemed to take from seven to ten hours to charge the plates, while it is not possible for the car to make over 35 miles a day. There were no grades, but some sharp curves. We thought that considering the expense and getting only 30 miles out of the battery, it was not much of a commercial success."

Mr. Everett states: "Almost all street railway men admit that the storage system, if successful, would be the ideal system, and all hope for its ultimate achievement; and in this age of progress it would be very short-sighted and bigoted to say that it will never come."

Descriptive.

Barking Road (London).—This line was started about the beginning of the year 1890, and has been in regular service since. The contract with the parent company was to run the cars at 4½d. (9 cts.) per car mile, driver included. The generating station cost £3,500; 6 cars at £450 each (batteries £300 per car); during the first 12 months 5 cars ran with regularity which left nothing to be desired. The road is fairly level and considered in good repair. The cars weigh 3.275 tons; the motors and gear, 1.36 tons; the batteries, 2.4 tons (22.6 per cent. of the total weight, loaded); the passengers, 3.6 tons; the total rolling load, 10.63 tons, of which 34 per cent. is paying and 23 per cent. is battery.

The following figures were supplied by Mr. Frazer, one of the contracting company's engineers: Number of car miles run, 100,844; running expenses, £3,390; wages at generating station, £677; wages of drivers, £602; fuel, £434; oil, £64; balance, £1,613 (not £1,777, as was reported), for depreciation and repairs. The battery depreciation and repairs amounted to £1,184, or 66 per cent. of the prime cost, which alone added 2.8d. to the cost per car mile.

The motor repairs are given as £1,000, or .428d. per car mile; this first figure (1,000) must be a mistake, as it does not check with the total. Assuming the latter (.428) to be correct, the motor repairs were £178, leaving £251 for other expenses. The item of depreciation of the electric plant on the car alone amounted to nearly 3½d. per car mile. The mere haulage expenses amounted to within 2d. of the total cost of tramway working in London, and have exceeded the cost of horse haulage by about the same amount. The result of the first year's working has been a loss of nearly £1,500; the running expenses per mile having been 8d. instead of 4½d. The figures regarding the subdivision of the items of expenses will be found tabulated and reduced to comparative figures under the heading "Cost of Operating," where they are compared with those of other roads.

Other Accumulator Plants.—The following data were published about some of the accumulator roads now running:

Hague (Holland).—Six accumulator cars are now running from the Hague to the Casino at Scheven-ing, a distance of about 3 miles. The speed of running is 12 miles an hour, including stops. The loaded car weighs 16 tons; it is 32 feet long, carries 68 passengers, and the battery of accumulators weighs 4 tons. The cars, constructed at Harlem, have two bogies of two axles each. Only one bogie is driven, having its wheels coupled together. The axles are connected to the motor by solid gearing, and the whole weight is carried by the axles. The motor is supplied by carbon brushes from a battery of 192 Julien accumulators, weighing 40 pounds each. This battery, when charged, provides current for a run of 45 miles, after which the cars return for change of cells. The accumulators are arranged in eight boxes or drawers, weighing half a ton each, placed under the seats. The car is provided with switches and resistances to allow the speed to be varied. The changing of the sets of cells is carried out by the aid of a special mechanical arrangement to allow rapid handling. Two sets of turntables and rolling carriers are placed at the end of a charging bench. The drawers, when taken from the car, slide easily on rollers placed on the bench and turntables, the operation of changing requiring five minutes. The spare sets of cells are always being charged as the others are being used. The connections are made with spring contacts arranged so that no mistake can occur. It is intended to settle the question of maintenance by the erection of a small manufactory of battery plates on the spot, with a laboratory.

Lyons (France).—The car is fitted with a motor by Alroth, of Basle, the current being supplied by 112 Faure-Sellon-Volckmar cells weighing $2\frac{1}{4}$ tons, the gearing being a Gall steel chain. The cells are placed under the seats and in two cupboards at the end, and the motor is placed beneath the car floor, so that nothing is seen but the switch handle alongside that of the brake. The total weight of the loaded car may be analyzed thus (giving the weights in hundredweights): Car 104, motor $13\frac{1}{2}$, switch and resistance 2, 40 passengers 53, battery 47; total, about 11 tons. The accumulators can be connected in four different ways, corresponding with 50, 100, 100 and 200 volts for 80, 40, 20 and 20 ampères respectively. The various couplings are brought about by a set of brush contacts under the car, moved by a handle with dial and figures. At normal speed, 7½ miles an hour (12 kilometres) on the level, the current is used at 10 ampères, rising to 30 and 45 on gradients. The capacity of the battery is 150 ampère hours, so that a minimum run of eight hours can be achieved without recharge. The lighting of the car is taken off the cells at 50 volts. So far the experiment appears to have been very successful, and the director seems strongly inclined to increase the number of electric cars.

Siemens & Halske road at the Frankfort Exhibition.—The car had two motors, each from 8 to 10 horse-power; they are very easily replaced in case of any accident or repairs. There are 162 cells of batteries of the Tudor system (a modified Planté cell), weighing about 2,200 pounds. The weight of the car without passengers is said to be about 17,600 pounds, and it carries about 40 passengers. The cells are used in three parallel sets of 54 cells (of about 110 volts) each; for starting they use a smaller number of cells, and in this respect it is an imperfect system, as the cells are thereby unevenly discharged. It takes 45 to 48 ampères to start the car. They will run for about five and one-half hours. The brushes are of carbon as usual. This system is not in practical use at present, but they are experimenting with it in Budapest.

Dubuque (Ia.) Storage Battery Railway.—A full description of this road, which has now been changed to a trolley road, will be found on page 61 of the issue of *The Electrical World* for July 25, 1891.

CHAPTER VIII.

UNDERGROUND (TUNNEL) SYSTEM.

As the requirements and the engineering features of an electric road running in underground tunnels are so very different from those of surface lines, they have here been placed under a separate heading. The contact wire, feeders, grades, curves, and many other features, are entirely different, and in many cases much simpler than in surface roads. The size of the units, the speed and the less frequent stops, particularly when compared to surface roads in a city with dense traffic, are all much more favorable than in surface roads.

There is, perhaps, no branch in electric railway engineering that is of such importance at the present time, and that has such a good immediate future as electric underground roads. The first road of this kind which was built, and which was to a certain extent an experiment, has proven so successful that the same parties soon after contemplated a second and larger line, which was followed very rapidly by propositions for similar lines in Paris, New York, and a third road in London. That such a system has no rival, is evident from the fact that steam roads of a similar kind, such as have been running in London for many years, have proven to be unsatisfactory as far as comfort to passengers is concerned, as well as in other features. The fact has already been demonstrated that individual motors are much more economical than individual steam engines, as they derive their power from a central station where large steam engines can be run to a much greater advantage than the small inefficient engines on locomotives. There are numerous other evident advantages, such, for instance, as the fact that the air is not vitiated by the smoke and products of combustion, including obnoxious sulphurous vapor, that the tunnel may be made much smaller, that the rolling · stock, and consequently the roadbed, may be made much lighter, and that the train may be controlled entirely from stationary block stations, instead of relying on signals, which, as was shown in a recent fatal accident in New York City, cannot be relied upon in the smoky atmosphere of a tunnel for a steam road. The units may also be made smaller, and the intervals between trains therefore more frequent than would be economical in a steam road. We believe that it is not over sanguine to state that before the next year is over extensive roads on this plan will not only be contemplated, but perhaps also in course of construction.

Mr. F. L. Pope, in referring to the underground tunnel systems, terms them "the most important development of electric transportation which has yet come before the public." Referring to the London road and to a contemplated similar road in New York, he states: "This undertaking having been placed in the hands of electrical engineers whose competency cannot be questioned, and backed by ample capital, has proved in London a phenomenal

success.   We are now assured that a syndicate of
the most prominent capitalists of New York have
undertaken to establish a system of transportation
upon the same plan, consisting of a network of tunnels at a depth of 100 feet beneath the surface, and
connecting all important points not only in the city
itself, but in the adjacent sections of Long Island
and New Jersey.   In such a system it necessarily
follows that the train must be run by electricity to
the exclusion of any other power.

"Statistics show that the traffic of the horse cars
and elevated lines in New York has increased 46
per cent. within the five years from 1884 to 1889.   A
careful estimate which has recently been made of
the movement of passengers in and about New York
during the year 1890 gives the following amazing
result:

New York city (surface and elevated roads)......................... 400,000,000
Brooklyn Bridge.. 38,000,000
Long Island ferries... 90,000,000
Staten Island and New Jersey ferries............................. 85,000.000
                                                                  _____
    Total... 603,000,000

"If the profit derived from carrying these passengers amounts to only one cent each per trip, it
nevertheless figures up to the snug little sum of
$6,000,000 per annum, which makes a very com-
fortable dividend upon a capital of $100,000,000.
Such a system would admit of the running of solid
trains at very short invervals through the city, to
and from all points in the suburbs, in every direction, and I venture here and now to predict that
within the next 20 years, if not the next 10 years,
electrical underground transportation will be
brought to such a state of perfection that a passen-

ger entering one of the central stations of New York may be deposited at his home station at any point within a radius of 10 miles in 15 minutes' time at a fare of five cents. And the magnificent result is to be the contribution to the convenience and the prosperity of the public of the electrical engineer."

Referring to the requirements of an electric underground system, Mr. Stephen D. Field stated: "In such an underground installation, using numerous power stations, very high potential, say 1,500 volts or more, can be safely employed, with consequent economy of operation. There is one very important point on which the comfort of travel materially depends. Great care should be taken to deaden vibration caused by the movement of the trains; continuous rails should be used, which, together with composition wheels and upholstered carriages, will do much to further this result. The locomotives should be 'direct coupled,' and in no case should the armature be placed direct on the axle without the intervention of springs. The power stations should be located at the middle and at either end of the line. Two insulated conductors would lead from these stations along the top of the tunnel. The conductors should be of soft iron, suspended from the tunnel roof in such a manner as to be readily accessible for repairs and cleaning of insulators. The insulators should be very strong and of great electrical resistance. Contact with the power conductors could be best secured by means of a 'magnetic trolley,' which gives great adhesive contact without much weight or resistance to rotation. Greater economy of operation will be secured if one large tunnel be used for two lines of tracks instead of a single tunnel for each. In the first case the air displaced by the moving train can find room for circulation, while in the latter it must be almost wholly forced out by each train. At moderately high speeds this becomes a factor of great resistance. In either case, however, plenty of air movement for thorough ventilation will be secured by the passage of the trains."

Further discussions on underground roads, which space prevents us from reprinting here, may be found in *The Electrical World*, page 247, March 28; page 196, March 7; page 402, May 30; page 437, June 13.

Descriptive.

City and South London Underground Railroad.— Owing to the impossibility of getting reliable technical data and detailed descriptions, the following short description and criticism has to be limited to such parts as could be seen from personal inspection, and to data obtained directly and indirectly from various sources. The officials, though courteous, evaded or declined to answer questions regarding details.

This railroad is the first and at present the only underground electrical railroad running through a large city for furnishing rapid transit between points in the city.' The road starts at Monument station, not far from London Bridge, which is near the busiest part of London, crosses under the River Thames, and continues in a slightly winding course under the streets, entirely through built up portions

of the city, to the district called Stockwell. Its course
forms to a certain extent a sort of diameter to the
circular course of the local steam railroad lines.
The tunnel itself consists of two lines of circular
subways, one for the up and one for the down
trains, sometimes located side by side, and sometimes over each other.   The tunnels are over each
other at various places, partly to avoid encroaching
on private property, and partly so that at the stations one elevator may be used for both lines, instead
of two, one on each side of the street.   It is claimed
that the cost of excavating and lining is less for two
small tunnels than for one large one.   The tunnels
are made of a circular steel cylinder 10 feet 6 inches
in diameter inside of the flanges and formed of
rings of cast iron segments bolted together at their
internal flanges. The rings are 1 foot 7 inches long.
The circular joints are made with tarred rope and
cement, and the longitudinal joints with thin strips
of pine wood. The surfaces are not planed, but
remain just as they came from the foundry.   The
tunnels pass for the most part through clay, and
partly through sand and gravel.   The tunnel was
driven by hydraulic cutters,* the whole being done
without the slightest interference with the traffic
along any of the thoroughfares under which it was
being built, a matter of great importance to the
residents of the city.   The line of the two tunnels
as they pass under the River Thames is not by any
means a level one. The greatest up grade is 1 in 30,
at the river, but otherwise the road is practically

---

* For a detailed description see *The Engineer*, London, June 7, 1889, p. 477,
third column.

level. The north end of the road, which is near the river, is about 80 to 90 feet below the street, but the average depth is much less than this, though it is said to be deep enough to pass under all sewers and pipes.

At each station there is a large hydraulic elevator, the power for which is supplied by pipes from the main power station, which is at the Stockwell end. Why such a cumbersome method was used instead of electricity is difficult to see; perhaps it was because the original intention was to run the trains by cable; perhaps also because the promoters were civil and not electrical engineers, who were not sufficiently well posted on the advantages of electrical elevators, and therefore resorted to the older and more primitive method. If electricity can be relied upon to run the trains, why should it be less reliable for the elevators? Should the trains stop by failure of the line, what is the use of the elevators, as there is a staircase next to each of them? To transmit hydraulic power through such distances must surely be more expensive and less reliable than to transmit electric power. We are informed that certain bad stoppages, which are made so much of by the opposition parties, have been due to the failure of the hydraulic transmission, which fact speaks for itself. The hydraulic plant at the generating station forms no small part of the total plant. As to its efficiency, we have no data, but it certainly cannot be as good as that of electricity would be. It would seem that the matter of transmission of power ought in a well-regulated plant be put into the hands of electrical rather than civil engineers,

To an American used to smoothly running railroad cars, the roadbed appears very rough, as noticed from the violent shaking of the cars, and this must affect materially the consumption of power and must strain the framework and joints of the tunnel. The air in the tunnel is cool and it is entirely free from the sulphurous odors in the underground steam roads, but it has a decidedly musty smell like that of a damp, deserted cellar. The cars are closed and there are no windows; the strong draught on the outside of the train is therefore not felt much in the cars. For long rides, however, this damp smell would be very objectionable, which could probably be overcome by building the tunnels so that they are drier. There is said to be only a small quantity of water leaking into the tunnels, which is pumped out daily by means of hydraulic injectors placed at the low points and operated from the high-pressure water pipes which operate the elevators. The underground stations are lined with white glazed brick, and appear to be dry and well ventilated; they are lighted by gas, strange to say. A train consists of one locomotive and three cars for about 30 to 35 passengers each, or about 100 to a train. They are run regularly at 4½ minute intervals, making a maximum carrying capacity of 1,330 passengers per hour. There are three trains running each way at one time about one mile apart, and one at each end station, making a total of eight trains, six of which are always running. The maximum speed is said to be 25 miles and the mean 20. The average speed, including stops, is said to be 15 miles per hour. The total length of the line is about 3¼ miles. The cars are nearly circular in section, have two rows of longitudinal side seats, and entrances at the ends; the entrances are closed and locked while the train is in motion. The cars communicate, and in case of an accident completely stopping the train, the passengers could walk out through the tunnel, provided the doors, which are locked and bolted from the outside, can be forced open by the imprisoned passengers. The cars are calculated for 30 cubic feet of capacity per passenger (20 cubic feet is the usual number required on English railroads). The cars are very inadequately lighted by incandescent lights, five in series on the 500-volt mains supplying the motor. They are naturally very unsteady, which is very annoying; all grades and changes of speed are very noticeable in the brightness of the lights. Oil lamps are used as reserves, and contrast very favorably with their poor competitor.

The locomotive has two axles, each with two wheels. Each axle carries a Gramme ring armature secured on it, and forms, therefore, a gearless motor. The magnets, one set for each armature, are formed of two large massive cores and a yoke piece, which extends from the shaft upward, in an inclined position toward the middle of the car; their lower end rests on bearings on the axles and the upper end is suspended by means of a bolt; each axle, therefore, has four bearings, two for the car and two for the magnets. Each motor is said to have a maximum output of about 50 horse-power, making 100 horse-power per train; when the train is fully loaded with 100 passengers, this would make about 1 horse-power per passenger, which seems very high; it is likely, however, that the normal power used is far below this figure. Each motor is said to be capable of running the train alone, though slowly, in case of an accident to the other motor, which case has, it is said, happened more than once. The reserve motor power is, therefore, 50 per cent. of the maximum, or 100 per cent. of the normal if horse-power can be taken as the normal. Another statement made to the writer by the attendant at the generating station was that each motor took about 75 ampères, which at 450 volts and 80 per cent. efficiency represents 36 horse-power at the wheels per motor, or 72 per car. There are two sets of carbon brushes held in tangential holders; the whole space surrounding the brushes and commutator is dust-proof and is covered with a thick plate of glass forming part of the floor of the locomotive, and thus permitting the engineer to see the action of the brushes all the time—a commendable feature. The brakes used are the usual air brakes; a supply of compressed air at 80 pounds pressure is carried in iron cylinders sufficient for one return trip. They claim that air brakes are much simpler than electrical brakes; besides this, the brakes might have to be used quickly and unexpectedly in case the current was interrupted, which itself may be sufficient reason for not using electrical brakes.

Regarding the electrical details and data of the locomotive and the circuits, no information could be obtained, although application was made in turn to the officials, the engineers, and contractors (Mather & Platt). But as far as could be seen, there was nothing unusual of any importance. Judging from this and from other things that were noticed, we surmise that the electrical part is not altogether satisfactory; in another similar road there would probably be changes made, as indeed is quite natural, because the present road is to a certain extent, at least, an experimental one, and it would have been surprising if this first road had left nothing to be improved upon. It would be quite interesting, however, for engineers to know what not to do. One thing we noticed was quite significant; in the repair shops there was seen a large scrap heap of copper wire, which, from its shape and general appearance, evidently came from burnt-out armature coils. An attendant at the repair shop remarked that they had become quite experienced armature winders, a remark which is significant, as the road had been running only seven months. The burning out of these armatures may have been due to the fact that they were originally poorly made, and perhaps the rewinding of them has been more satisfactory; at all events, the insulation of the wires of a gearless armature (that is, one whose shaft is the axle of the wheels of the car) is strained very badly, particularly when the roadbed is very rough, as in this case. No springs can relieve it of the direct blows which the armature gets when the track is not perfectly even. Besides this, the strong current required at starting is far greater, and acts for a longer time than in a geared motor. This is readily noticed from the variable brightness of the lamps in the car, which in this case can almost be said to be a crude ampère meter.

The locomotive receives its current from a single rail laid on the sleepers near one of the tracks; the tracks themselves formed the return lead. This current rail was made of an alloy of steel (the composition is said to be a secret) which is at the same time very hard and a good conductor. The rail has a rectangular section, and lies almost directly on the sleepers. The insulation appears to be quite thin and without much surface; it is likely, therefore, to cause trouble by leakage in such a damp place. Why such a good opportunity was lost to place it on the side or top of the tunnel is not apparent; this would have avoided almost completely the difficulties of insulation. The contact from the locomotive to this rail was made by a very crude arrangement consisting of a heavy cast-iron tongue which slid on the top of the rail and was held by a loosely fitting hinge; there was no provision for shunting the poor contact at this loose hinge, though it might easily have been done. There were three of these tongues on each locomotive; being curved, they work equally well for both directions of motion. The contact rail was in sections, supplied by independent leads, presumably the same as usual.

There was only one generating station, which was located at the Stockwell end, presumably because ground was cheaper there than in the middle of the road, the space occupied by this station and the adjoining repair shop being quite large. It included an inclined plane tunnel leading down to the subway for raising the cars for repairs; this was operated by a cable and a windlass. The main part of the interior of this station was shown in an illustration too large to be reproduced here, which will be found on page 301 of *The Electrical World*, October 24, 1891. There are three dynamos of 500 volts and about 500 ampères each, having a speed of 500 revoluitons. There is an engine of 375 horse-power for each; they are non-condensing, and govern very poorly. The dynamos are compound wound, and contain switches on their yoke pieces, by which the compound winding on each limb of the magnets may be cut out. There is no equalizing arrangement between the compound windings. Link belts and idle pulleys are used. The belts stretch very materially when carrying a load. The dynamos are cooled artificially by air blasts supplied from air pumps. The lubrication is by means of a milky white mixture of castor oil and water.

The switchboard for this station is quite simple. The dynamos are connected separately to it, and can be variously connected to the different sections of the road. It includes an ampère meter to five ampères, by means of which the insulation is measured every morning before starting with a 500-volt current. We understand that a few ampères leakage is not considered to be too much. At 2½ ampères the insulation resistance would be but 200 ohms, which surely is not good. The cost of this wasted energy (.6 horse-power per ampère of leakage) is probably not of as much importance as the electrolytic corrosion of the rails and destruction of the insulation which probably accompany it. The switchboard, furthermore, contains automatic cutouts, which for abnormal currents switch a resistance into the circuit; if after that the current is still too great it opens the circuit entirely; if the latter does not take place the engineer in charge cuts out the resistance again.

Strange to say, there is no telephonic communication between the generating station and the way stations. Instead of this there is a needle telegraph system. The reason given for this otherwise poor substitute is that all communications must be made in writing and filed at the office, which naturally has its advantages.

Inquiry at the main telegraph office of London showed that there were no serious disturbances of the telegraph lines by these powerful earth return currents; but they are said to materially affect the magnetic observations at the Greenwich Observatory, which is about five miles from this railroad. The nearest earth plate to the railway is $2\frac{1}{2}$ miles distant, but in spite of this the currents are regularly and continuously visible from 7 in the morning to 11 o'clock at night. The differences of potential registered are, of course, comparatively slight, ranging, however, from a small fraction of a volt to nearly half a volt.

The interruptions in the running of this railroad are said to have been due chiefly to the hydraulic pipes and to short circuiting and burning out of the armatures. But these interruptions are said not to have been of a serious nature. The former could readily have been avoided by using electric instead of hydraulic transmission of power. For the latter electricians will soon find a remedy, be it by more careful insulation or by using a geared motor which is supported on springs to dampen the concussions.

Much is made by opposition parties of the interruptions which have occurred, but this is presumably exaggerated. In the first 11,000 trains run, there were only 18 failures. We surmise, however, that there was, perhaps, too much civil engineering and too little electric engineering in some features of the road, and we are led to believe from this that the latter engineers, who were mere contractors under the former, were not altogether responsible for some of the unsatisfactory features.

The road was opened in December, 1890. The uniform fare is 2d. (4 cts.). During the first 27 working days 414,000 passengers were carried. Up to February 25th, the earnings were £7,500; 900,000 passengers were carried, 60,000 train miles were run. The receipts per train mile were 2½s. (60 cts., or 20 cts. per car mile for trains of three cars). Reports in April state that there were about 100,000 passengers per week. During the first six months 2,412,343 passengers were carried; 141,408 train miles were run; gross earnings, £19,688; receipts per train mile were therefore 2s. 9d. (66.8 cts., or 22.4 cts. per car mile); total expenses, £15,520, or 78.8 per cent. of gross earnings; net earnings, £4,168, which was only sufficient to pay an ordinary interest on the bonds. The running expenses per car mile were 11d. (22 cts.), from the above figures it was 26.4d. (or 52.7 cts.) per train mile, (or 17.6 cts. per car mile for trains of three cars); average number of passengers per train mile, 17, that of the Metropolitan Steam Road being 43. Average receipts from passengers, 2s. 9d. per train mile. For October the average weekly receipts were said to be £750; operating expenses were 3.08 cts. per passenger. The fare during the busy hours is to be increased from 2d. to 3d.

Central London Underground Railroad (Projected).—The best proof that the present City and South London road described above is a success is shown by the fact that this second similar road is projected by the same parties. This new road is to extend between Shepherd's Bush and the Royal Exchange, a distance of about six miles, under Cheapside, Holborn Viaduct, Oxford street, etc., which is the main artery running east and west through the heart of London. There is said to be more traffic there than in any other thoroughfare in the world, Broadway in New York included. This route forms a sort of diameter of the circular route made by the present underground steam road. As a line of travel it is very similar to Broadway in New York.

The tunnel will be chiefly under streets, at an average depth of 60 to 70 feet, at times reaching 80 to 90 feet, passing under the present underground steam road in two places. The steepest grade will be one per cent. The tunnel will be 11 feet 6 inches in diameter, which is larger than that of the City and South London road by 18 inches. This will enable the cars to be larger and more comfortable. The cars are cylindrical, fitting the tunnel quite closely. A train will have four cars, carrying 350 passengers (seated), and will run on three minute time, instead of 100 passengers on four and a half minute time, as on the other road. The present road does not pass through a district where there is as much traffic as in the line of this new road.

The average speed will be 15 miles an hour, the total rolling load 120 tons. There will be 13 stations. The generating station will have from two to three thousand horse-power. The proposed capital is £3,500,000. The cost is estimated at about £600,000 per mile, and the working expenses 9d. per train mile.

It is very significant and instructive to notice what changes in the construction are to be made in this new road as compared with the present one, the promoters being the same. Any one interested in such roads will do well to compare the two roads in detail, when such detailed descriptions are published. At present such information is scarce. It appears that the general features will remain about the same. The feature of having two independent motors on the locomotive, one for each pair of driving wheels, is said to be very satisfactory; one of these motors is said to be able to run the train alone in case the other is disabled. One of the principal changes is that there will be a separate system of mains for the lighting of the stations and tunnels, and probably the cars also, so as to avoid the very annoying and unceasing changing of the brightness of the lights. The brakes will be operated by compressed air carried on the locomotive, and supplied at the end stations as at present; electric brakes are claimed to be so much more complicated than air brakes. The strange feature in the present road of running the large elevators at the stations by hydraulic pressure, transmitted through miles of pipe from the generating station, will probably be changed in the new road, in which electrical elevators will probably be used. To an American, so accustomed to electrical elevators, it seems strange that this was not done on the present road instead of making the system more complicated by the introduction of a second system of tramsmitting power which, it would appear, is more complicated and more liable to failure than the electrical transmission for which the circuits exist. If the latter is relied upon for propelling the trains, why should it be less reliable for running the elevators? Hydraulic transmission surely cannot be cheaper.

Paris Underground Railway (Projected.)—It is reported that the municipal authorities of the city of Paris have just decided on building an underground electric railroad from one side of the city to the other, as a sort of a long diameter to the nearly circular city, passing near most of the prominent parts of the city. The system adopted is that presented by M. J. B. Berlier. The concession includes an underground tunnel nearly following the course of the Seine, and traversing Paris at its greatest breadth. There are to be six stations, and a generating depot at the end of the line. The construction of the tunnel will be very similar to that of the City and South London underground electric railroad.

The transversal sleepers of injected wood are placed a metre apart; on these run two parallel ways formed of rails 30 kilos to the running metre. In the middle of each line is placed a central rail insulated by india-rubber plates from the sleepers to which it is fastened. The motor carriage carries under its framework two gearless motors on the same circuit. The motors are each 25 horse-power, and will be regulated for proportionate speeds and for stoppages by a rheostat regulator, while the reversing of the machinery will be effected by a special commutator. During the journey, all the doors will be closed by an electric bolt under the charge of the engine driver; on arriving at a station they will be released and will open automatically. The lighting question is to be solved by a special service established at the works at the end of the line, and will include arc lighting for the stations and incandescent lights for the carriages, signals, and for the general service of the tunnel throughout its length. The electricity generating machinery will be duplicated, so as to provide against breakdowns.

A Four Track Tunnel Road Proposed for New York City.—In the supplement to *The Electrical World*, of February 21, 1891, there is published an interview of Mr. F. J. Sprague, on "The Solution of the Problem of Rapid Transit for New York City," of which we give below an abstract of the plan which he proposes.

The first portion of the interview is a consideration of the various systems proposed for New York, which we will omit here. He states that the total annual passenger traffic, that is, the total number of passengers carried in that city, has increased at the rate of 140 per cent. in each period of ten years, since 1866, and is now something over 325,000,000. At the same rate of increase, it would amount in 1890 to over 500,000,000, and in 1900 to 1,225,000,000.

He states that a road should be built not only in the most substantial manner, but with well-hedged privileges and franchises, which will protect the city in its undoubted rights, and offer the certainty of fair and equitable returns to an investor, large or small, so that its bonds or stocks should be as safe and as much sought after as a government bond.

The following is an abstract of the description of the plan which he proposes: "An independent and way express service, each having double tracks, but having independent paths, run in independent circular iron tunnels, and forming a loop around the city from the Battery to Jerome Park; the express and way tracks to intersect or join at different levels at common stations, and to be operated by electricity; way tracks to have stations every third of a mile, express every mile and a half; and a loop to be provided just south of the Harlem River. Independence of express and way service is essential. The operation of express and way trains on a two-track railroad, to meet the condition of service of New York City, should not be thought of. It is impossible to satisfactorily so operate to-day; it will be vastly more so twenty years from now.

"To illustrate this point, we have only to consider the work done on the Third avenue elevated road, such as it was a short time ago, and it cannot be less to-day. I have made a special study of it for six years past.

"This road is eight and a half miles long. Grades vary from 8 to 105 feet to the mile. The level stretches amount to about one-third of the whole distance, and this includes the stations. On the 17 miles of single track there are 52 stations. In the busy hours there are no less than 63 four and five-car trains on the track. These trains, weighing from 80 to 95 tons, make the half trip in 42 minutes, including stopping at 26 stations, or at the rate of about 10 traffic miles per hour. The work of the engines may be divided into three parts, viz., overcoming the train's inertia, lifting the train on the grades, and traction, and the maximum is at least seven times that necessary for traction at mean speed on a level. Three times in every mile this weight of 80 to 95 tons must be started from a dead rest, raised to a speed of 20 or more miles an hour and brought to rest in about 80 seconds. The engines are run with 130 pounds boiler pressure and have a capacity of 185 horse-power. Fifty-nine per cent. of the power on a round trip is used in accelerating speed 24 per cent. in lifting and 17 per cent. in traction, and the average power developed per minute per round trip, including stoppages, is 70.3 horse-power.

"These are instructive figures, which alone should show the fallacy of a two-track system. They show that it is perfectly certain that on the present system a greater traffic speed than 10 miles an hour cannot be maintained, while it is equally true, on the other hand, that we cannot consider any less frequent stations for way service than three to a mile, yet note what tremendous duty is required of engines to make even this moderate speed under conditions of frequent stops. I am well within bounds when I say that a motor which can do the service of a way train at short intervals can equally well do the work of a heavy express train at long intervals. If an individual tunnel construction be used, there would not only be no room for a train to fall, but a broken axle or a derailed train would not, of itself, be apt to prove serious. The tunnels would be lighted constantly and uniformly; the rails would be always in perfect condition for adhesion, and the temperature of the tunnel would be almost constant. Heating of the trains or stations would be unnecessary. Ventilation could be made perfect.

"Cars for such underground systems would require no windows, landscape views not being essential to rapid transit in a city, and hence, with the same weight, can be more strongly constructed. Absolute freedom from sharp curves and the arranging of grades so that, instead of being a detriment, they can be of positive aid in train service, are not among the least advantages of an underground system. Where curves occur, they can be either on a dead level or the grades can be so managed that they will be of service.

"Having determined upon a tunnel construction, independence of express and way tracks is not all that is essential. These might be obtained if all four tracks were in a common tunnel, but when we realize some of the other conditions which must be met, and that such a four-track system could not be placed in a common tunnel not less than 50 feet in width and 20 feet in height, objections to such construction are pronounced. It is manifest, also, that if the size of a tunnel for four tracks has objections structurally and otherwise, one for two tracks must have similar objections, although in a less degree. Hence my plan is a system of independent express and way routes, each having two tracks, each carried in independent tunnels, and having their paths intersect at certain common important points, say on an average every mile and a half. I would have not only the routes of these services independent of each other, save at their common meeting points, but I would have them, at all other points, independent also in the matter of grade, and furthermore, I should have the tunnels of each system, while normally running side by side, sufficiently independent to permit of such variation of route for each track as the meeting of obstructions in the construction should require; or briefly, I would have four independent tunnels, all four meeting at a common station once every mile and a half, and the way tunnels meeting at common stations every third of a mile.

"The advantages of such a method are: 1. A much stronger construction is possible with the same weight of materials, because a tunnel of 12 or 13 feet in diameter can be built with strength to support a given outside pressure with much less than one-quarter of the weight of material required to build a tunnel of four times the area, and a tunnel to accommodate four tracks would have more than four times the area of a tunnel to accommodate one.

"2. At any given depth there will be far less interference with and danger to the foundations of a building with a small tunnel than with a large one; in fact, a tunnel of the smaller size could be run with perfect safety almost through and certainly under or alongside the very foundations of the heaviest buildings,

"3. The nearer the surface a system is run the greater the variety of obstructions which will be met, which with a large tunnel would require a very costly change of gas, sewer, and other pipes. A 13-foot tunnel, independently run, could easily deviate up or down, to the right or left, in fact, weave its way through and around obstructions, which it would be impossible to do with a large tunnel. This is one of the very greatest advantages to be obtained with the independent tunnel construction.

"The express route, if a difference of level exists between express and way tracks, should be the lower one, and being lower, it can run in more direct lines with less necessity of divergence on account of obstructions. While express tracks should thus run in the most direct line possible between main stations, the way tracks could take more or less divergent paths as determined by the best positions for the way stations. On this account the express route will be somewhat shorter than the way, which would be an important help to a quick schedule time.

"If, as should be the case, the tunnel is built of iron and of circular cross section, not only will the construction be the strongest which can be built for any given clearance, but with the minimum of excavations. Its safety would consist in its arch presenting the maximum of resistance to vertical and side pressure. The thickness can be varied to suit the condition of strain to be met. The entire New York City line could be ready for active operation within 18 months. The cost of such a construction would be about $500,000 per mile of single tunnel. This facility of access to trains can only be had where there is a common station, which station should be between the up and down tracks, and should be roomy, well-lighted, and easy to reach from the street. Sole dependence should not be placed on elevator service, although such should be provided whenever their use could be of advantage. At way stations a common platform between two tracks is easily provided, because the two tracks are on the same level, and there is nothing to provide for special changes. At the common way and express stations, however, a somewhat more elaborate station system should be adopted; the depth of the way tracks below the surface would be the same here as at regular way stations, but the express tracks would be about 13 feet lower, and preferably directly under the way tracks; thus the four trains would occupy the four corners of a parallelogram. Elevators, when used, should be in the centre of the station, open on both sides, and operated primarily by electricity. The entrance to stations should take no street room, and hence form no obstruction. The station and offices should occupy the first floor and deep basement of a building especially constructed for the purpose, all the upper portion of which could be utilized for ordinary business purposes, and two or more elevators could run the full height of the building from the station platform.

"There should be no grade crossings. While the importance of this rule has met little recognition in this country, and does not mark any widespread practice here, yet it is the standard practice of the main trunk lines in London. Its merits need little comment, because whatever other conditions may exist, high speed, safety, and regularity of service can only be had in a fully satisfactory manner where there is a clear right of way.

"There should be no connection between an express and a way track, except at the dispatching and siding stations, where everything could be under the most rigid personal supervision, and where the necessary complication of tracks can be properly supervised. Switches, even between the way tracks and between the express tracks, should be used only in case of emergency, and the main line of rails should be absolutely unbroken. The roadbed should be of the most rigid and even character, and the weight of rail commensurate with the duty.

"Train intervals and passenger requirements should determine the length of trains, instead of length of trains determining train intervals. It is not express service to have to wait 10 or 15 minutes for an express train after leaving a way one. The most frequent interval that can be safely run is that which should be adopted, of course, with reasonable regard to the amount of passenger traffic, which, however, will be better accommodated at certain portions of the day by smaller units and more frequent trains. The number of cars in a unit determines the number of passengers that can be carried with a given safe interval between trains, because the time of discharging and taking on passengers will be almost constant with a given traffic per car, whether the trains be long or short. So

long only as there is a safe interval between a train
standing at the station and one approaching it, the
train unit is not too small for passenger requirements, providing the following train is not delayed.
When, however, the following train has to wait
on approaching a station for another train to clear
the way, the train unit is too small. The use of way
and express tracks would relieve that congestion of
traffic which characterizes the two track system of
the elevated railroads, and which on any two-track
road cannot be avoided in New York.

"Electricity will unquestionably be the motive
power. The hum of the motor is the song of eman-
cipation. As an independent way and express system of tracks should be an essential of a rapid transit system, so also should the use of electricity as a
motive power be a *sine qua non,* and accepting the
statement that this agent is capable of satisfying in
the highest degree the most exacting demands of
service, the system should be planned with a special
reference to its use. One great trouble with most of
the rapid transit plans which have been proposed is
that they have been designed from a steam
engineer's point of view. Steam practice has determined not only the form of roadbed or tunnel, but
also the laying of tracks, the construction of the
cars, the ventilation and lighting of the trains and
the roadway, the length of trains, the system of
switching and dispatching—in short, there has been
scant recognition of the fact that electricity has,
within the past three years, jumped to the front
with tremendous strides, and given practical and
most conclusive proof of its pre-eminent fitness for

at least all transit of this character. Many of these systems were planned before electricity was in its swaddling clothes, and even recently when electricity has been spoken of by the steam engineer, it has been with a somewhat indistinct idea that electrical engineers would fit their conceptions to steam demands. But this will not be done. The best informed of them know that the time has come when it will accept no secondary place; that there is nothing which steam can do in the matter of handling trains, certainly within the limits here considered, which electricity cannot perform in a more satisfactory and a more perfect manner.

"The method of supplying the current will undoubtedly be by the overhead single conductor underneath contact system, with rail and tunnel return, despite the criticisms of what is properly known as the overhead wire system, which is the necessary street development, with many limitations, of the underneath plan. With a tunnel road the overhead wire would be replaced by a rigid rail, supported at short intervals at a fixed distance from the roadbed, and following accurately the centre line of all track and switch paths.

"We would then have a system in which the electrical part, or rather the overhead conductor, because all parts would be electrical, would be just as rigid and permanent as the roadbed, and we should have perfection of insulation, freedom from personal liability, continuous contact with a single device, and speed limited only by the capacity of the motors. With such an overhead conductor, a speed of 120 miles per hour has already been successfully obtained on experimental work.

"The so-called difficulties of the overhead system would then absolutely disappear. A centre rail system, with the conductor near the level of the tracks, must necessarily be broken at all switches, thus requiring double contacts on the motor car, and cannot offer the same advantages of insulation and freedom from personal contact. The overhead system would have been used upon the London road if there had been room enough. Of this I was assured by the consulting engineer of the road. The current could be supplied to this entire system for not only the motive power, but for lighting, and, if necessary, for ventilation. The station would be operated by triple-expansion condensing engines in three units, each driving a pair of multipolar slow speed dynamos directly coupled, the units being large enough so that two engines could handle the normal maximum demand upon the station. The economy of coal consumption under these conditions is too well known to need comment.

"While for the purposes of traction, getting quickly under way, and for marvelous braking power in case of emergency, a motor on every truck with control from a pilot car is possible, and would be a great service, I think it likely that an independent motor car will be found to be more advantageous, because of the care which the machines can have, the lack of necessity to subordinate the motor to abnormal conditions, and because, with the grades which an underground system will present, and with the constant co-efficient of traction which the rails would have, a motor car could be relied upon to efficiently handle six times its own weight of passenger cars behind it. The multiplicity of parts should ordinarily be avoided, and the construction of the cars which the tunnel can best have might make the use of a motor under each and every car inconvenient.

"It is unnecessary to go into the question of motor construction. Gearing will be done away with, slow-speed motors used, directly coupled, and it is safe to say that they will be quite as reliable as the best steam locomotive.

"For use on an electric train, not only is every system of braking in common use, the vacuum, the air—whether the continuous pressure or the stored resevoir system—available, but more than this, and acting with a certainty, rapidity, and delicacy almost inconceivable, is the dynamo power of the motor system. In a perfect system of transmission, every car running on a down grade, while not accelerating its speed, and every car coming to a stop ought to be of service in helping to propel other cars.

".As already pointed out, the work done on a train may be divided into three classes: First, the work of accelerating the speed, that is, of getting from zero up to its maximum speed, the process of imparting to the train a certain amount of stored up energy which, when trains are stopped by the use of the brake in the ordinary manner, is nearly all wasted. Second, in lifting the train from one level to another, that is grade work; this is partially recovered on down grades, more so on express service than it can be on way service, on account of the less frequent stops. Third, traction, the work expended in hauling a train on a level.

"In a cable system running on a straight line, cars which are running on down grades do help to pull others on level or up grades, but the amount of power used to propel the cable itself is so largely in excess of that used for propelling the car that this is not of much importance, and on roads with curves it becomes even less so. Slowing down cars on cable roads is of no service in restoring energy to the system. In an electric car, however, when operated under proper conditions, all this is changed. There is intimate relationship between every foot of the system, and a train slowing down or a train running on a down grade can give back a large proportion of the energy to the system for use on other trains. The transmission of the energy by the current proceeds with calm indifference to grades and curves, or the condition of the thermometer or the barometer, and serenely annihilates time and space. Be it hot or cold, wet or dry, the interchange of energy is perfect.

"This method of braking is of great commercial importance, since by it there is a saving of about 40 per cent. in the amount of power at the central stations, and it effects a similar saving in the size and cost of stations and conductors. Instead of the current being all supplied from the main generating stations, it would be supplied from nearly as many moving stations along the line as there would be trains slowing down or running on a down grade. In fact, the loss of the two conversions and one transmission would be fully counterbalanced, and the original horse-power at the central station would be no more than the aggregate net horse-power developed on the entire line.

"Another method of using the motors for braking purposes is to break the connection with the line and reverse the machine through a local circuit on the train, varying either the local circuit or the circuits of the machine. While in this position of braking, if the local circuit is opened and the line connection made, the machine would be instantly reversed.

"Elaborate experiments have been made in handling a car without shoe-brakes by these methods of braking. When desired, the braking could be made so sudden as to cause the wheels to have a continuous skidding rotation, not such a skidding as is caused when an air brake is put on too sharply, but a rotating slip which is just enough to relieve the motor when the strain on it reaches the point determined by the co-efficient of adhesion on the rails. This is the ideal method of braking, because fixed skidding and flat wheels are an impossibility, and the wheels will turn until the train comes to a dead stop, although where the braking power is put on too suddenly and exceeds the grip of the wheels, they will relieve themselves by slipping just enough to keep the braking at the maximum limit. A train so governed can be made to creep down the maximum grade of a road at a snail's pace, and in an emergency such a car, running 21 miles an hour on a down grade, has been stopped and reversed within 90 feet.

"When an electric motor is reversed, its action is cumulative, and thus much increases the effort to stop itself; and the higher the velocity at which it is traveling the greater this effort. In fact, it would be practically impossible with a properly constructed railroad motor to prevent its reversing when the switch is thrown over far enough."

Reno Underground Road (Proposed).—Among

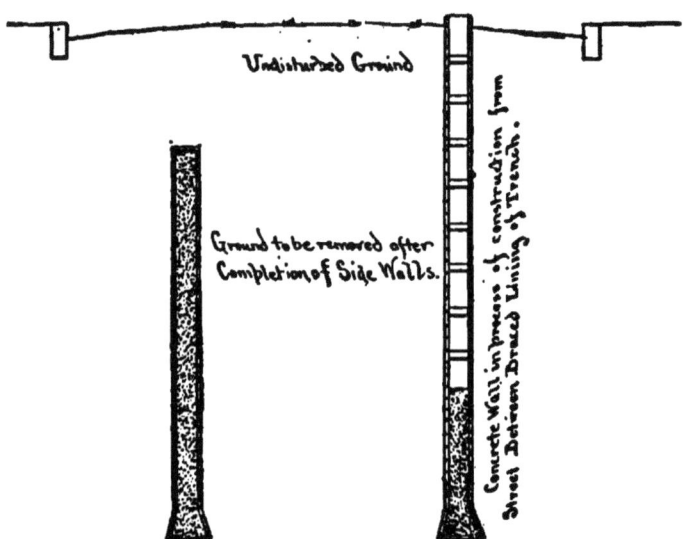

FIG. 26.—METHOD OF CONSTRUCTING THE RENO TUNNEL.

the proposals made for an underground road for New York City was the following, proposed by Mr. Reno, which contains some interesting suggestions.

The proposed method of construction is shown in Figs. 26 to 30. One side wall at a time is constructed. A trench outside of the car tracks is excavated 3 feet in width and 38 feet deep, and the ground thoroughly supported and braced by a plank lining.

The tunnel wall is then made by filling in this space with the best concrete. When the wall is built up to within 11 feet of the surface, the plank lining is removed, the trench filled in, and the strip of street paving relaid. The other side wall of the tunnel is similarly constructed. The excavation of the ground between the side walls is then effected without disturbing the street surface, pipes or subways, by a

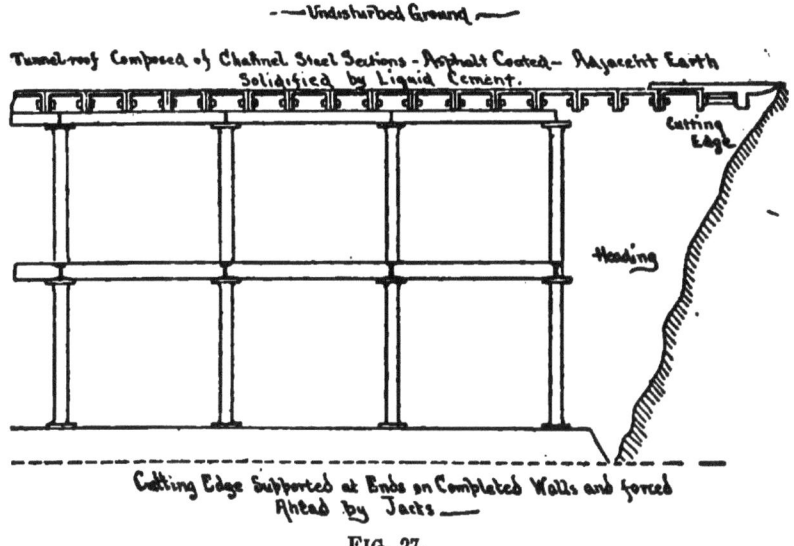

FIG. 27.

steel "cutting edge," as seen in Fig. 27, which rests at each end upon the completed side walls and is forced ahead by hydraulic pressure, while the channel steel roof sections are built in behind it and the construction of the internal steel structure immediately follows. The ground adjacent to the steel roof is solidified by liquid cement forced in under pressure. This in conjunction with the asphalt coated

steel roof is to insure a perfectly dry tunnel. The row
of iron columns 12 feet apart through the centre of the
tunnel support a longitudinal steel girder, upon
which rest the roof sections. These iron columns
are braced from the side by the transverse girders
that support the track beams.

The proposed method of construction will admit
of great rapidity in driving the tunnel, as the side

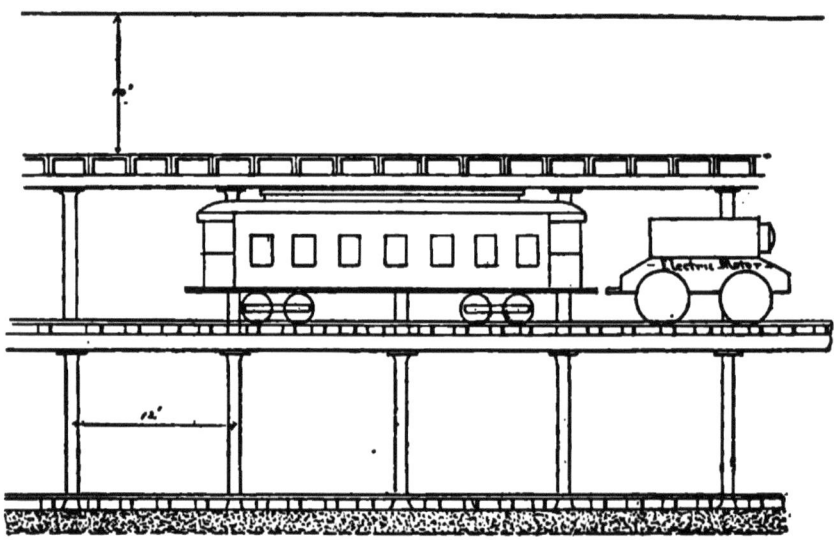

FIG. 28.—LONGITUDINAL SECTION OF THE RENO TUNNEL.

walls can be built in advance of the "heading," and
all four tracks could be utilized in removing excavated material. With five points of attack giving
ten "headings" and an average speed of 10 feet per
day each, the tunnel proper could be constructed
from Battery Park to Fifty-ninth street (five miles)
in 264 days. The building of the stations could be
carried on while the tunnel is in progress.

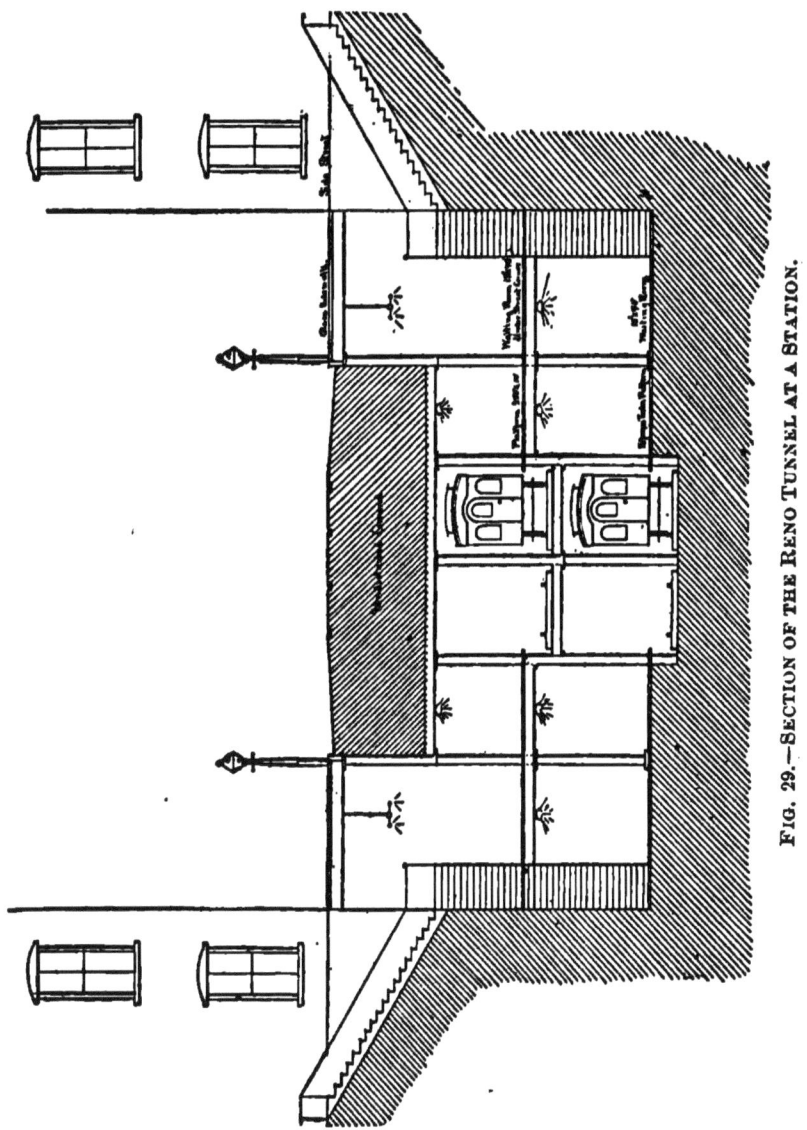

FIG. 29.—SECTION OF THE RENO TUNNEL AT A STATION.

The tunnel is constructed on the principle of "square sets" used in mine timbering, the horizontal girders resisting the pressure of the tunnel's side walls, while the vertical columns divide the span of the roof. The proposed method of constructing the roof and driving the tunnel by side channeling and small blasts insures rapid advancement in rock work, and will not unsettle the ground surrounding underground pipes and subways, nor interfere with the construction or operation of cable railways, nor will street traffic be seriously disturbed in constructing concrete walls from the surface. The outside width of the tunnel being only 23 feet, these side walls can be built without injury to the foundations of adjacent buildings.

The line of iron columns through the centre of the tunnel permits the use of a flat roof, which may come to within about 10 feet of the surface without disturbing underground pipes or subways. It will readily be seen that the open spaces between the girders and ties that separate the two levels of the tunnel allow free circulation of air, and with the use of electric motors insure good ventilation.

Three large power stations would be located along the Hudson River, where real estate and fuel are cheap and water for condensing is plentiful. In this way power could be generated at the rate of two pounds of coal per horse-power instead of 10 pounds per horse-power now required on the engines of the New York elevated roads. The electric conductor (which could be of heavy copper of T-shaped cross section in order to give it rigidity) might be sup-

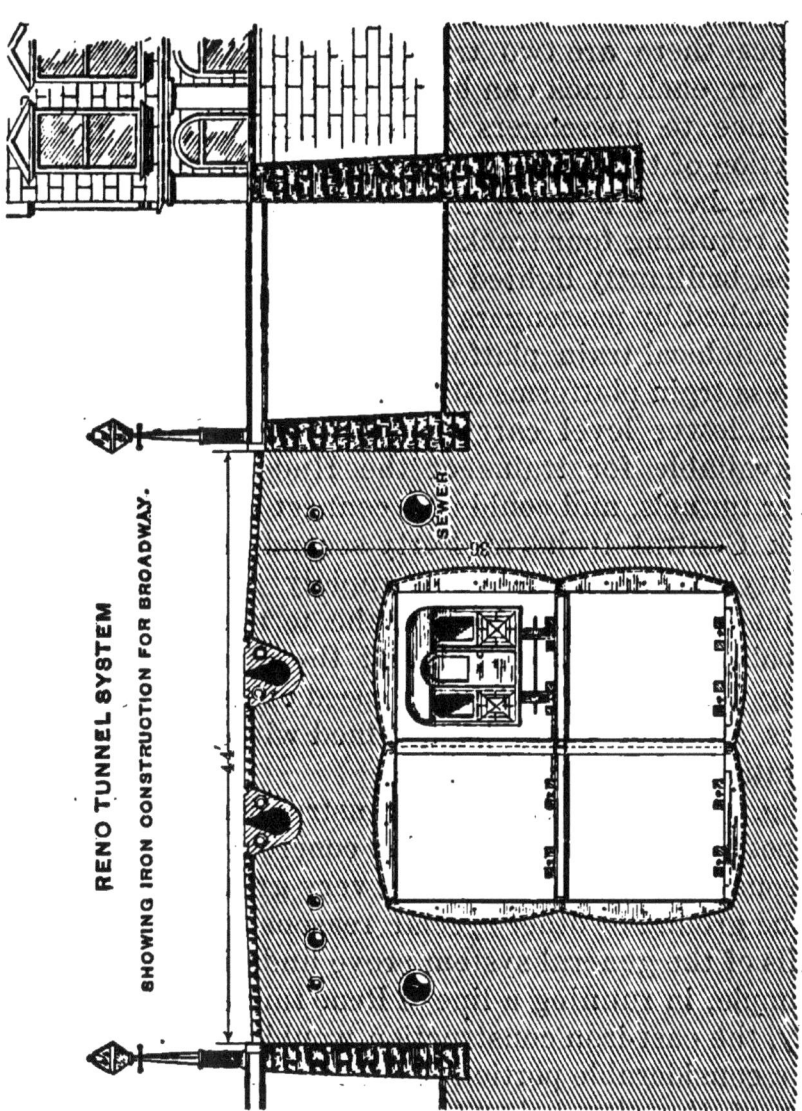

RENO TUNNEL SYSTEM

SHOWING IRON CONSTRUCTION FOR BROADWAY.

FIG. 30.

ported above the cars, using the rails and structure as an auxiliary conductor for the return current.

Since there are two tracks only, on each level, cars on each track can be reached without inconvenience to passengers or complicating the construction of the tunnel at the station platforms, this feature being a great advantage over tunnel systems requiring four tracks on a level. Station platforms, brilliantly lighted by electricity, could easily be reached by passengers from the surface, the distance to local train platforms being 20 feet, and to express train platforms 33 feet.

The lower level carrying express tracks would be available for trains of the Hudson and East River tunnels, and could also connect with the New York Central trains at Forty-second street. The facility for handling trains rapidly by electricity would be such that loops at the terminals would probably not be required, and the express tracks being on the lower level, a speed of as high as 40 miles an hour could be attained without causing vibration of the structure.

Fig. 30 shows a cast-iron construction that it is proposed to substitute for the concrete walls and masonry in places where, from very great dampness of the ground or other special reasons, the concrete walls of the general system prove undesirable. For example, in running a line of Reno tunnel up Broadway, the cast-iron construction would be employed for a considerable portion of the length in the lower part of the city, below Grand street, for instance. In this the walls, roofs, and floors are composed of flanged panels of cast iron about two feet in width, set in place in the rear of a shield pushed forward hydraulically; this shield will not be circular, as in most cases where it has been employed, but rectangular in section to cut the tunnel to the exact size necessary. The shield will be advanced for two feet, and then the eight iron panels will be placed in position and bolted at the flanges, after which a grouting of liquid cement will be forced through holes left for the purpose and form a solid packing that both prevents the entrance of water and shields the iron from corrosion. The method is a common one in tunneling work, and has been employed on the Croton Aqueduct. In the upper part of the city the plans described first would be followed as being simpler, cheaper, and more easily applied in ground of the character to be found on that part of Manhattan Island. As before mentioned, the grouting would be freely used to make the tunnel perfectly tight and prevent any difficulties from moisture or sewer gas forcing their way in.

The following estimate per mile (excluding the stations) furnished by Mr. J. W. Reno, the inventor of the plan, is based upon similar contracts in the Croton Aqueduct, and places the total at not far from one and a quarter millions per mile:

EARTHWORK AND MASONRY.

Outside area of tunnel, 24 × 28—allowing an average of one-half rock work and one-half earth work:

66,440 cubic yards earth excavation at $2.50		$166,100
66,440 " " rock " at $6		438,504
Side walls (concrete, 3 to 1), 2 feet thick at top, 3 feet at bottom and flaring out to 4 feet base; 28 feet high:		
Two walls, 28,160 cubic yards concrete at $6		168,960
Earth excavation for same, 17,600 cubic yards at $1.40		24,640
Rock " " " 17,600 " " at $2.50		44,000
Line of masoury under centre iron columns, 1,460 cubic yards at $10.		14,608
Concrete floor, 18 inches thick, 4,671 cubic yards at $5		23,355

STEEL STRUCTURE.

440 cast-iron columns, 24 feet long, 12 inches diameter, 2 inches thick = 4,883 pounds each, at 2 cents	42,680
21.120 lineal feet 12-inch I-beams, 781,440 pounds, at 3 cents	23,443
440 transverse 12-inch I-beams, 22 feet long each = 9,680 feet, at $1.11	10,744
5,280 feet 20-inch I-beams for centre of roof, at $2.20	11,721
5,280 feet "channel" steel sections for roof, 22 feet long, 12 × 8 × ¾ inch, 1,552 pounds each, at 3 cents	245,836
87 pounds steel rails, 14,080 yards = 612 tons, at $30	18,360
10,560 creosoted ties (2 feet apart), at 60 cents	6,336
Spikes, fastenings, etc	5,000
Total	$1,232,371

Munsie-Coles Underground System (Proposed).—In

FIG. 31.—SECTION OF THE MUNSIE-COLES UNDERGROUND SYSTEM.

its construction the transverse girders are made in sections, to admit of placing the vertical columns in position without obstruction to the daily street traffic, and, by a proper distribution of labor and material to make it possible to place the superstructure and new paving in place without any interference with the ordinary travel of the street. By two open slots from the tunnel to the surface of streets ventilation of the tunnel is provided for while the labor of excavating is in progress. The vertical depth required from the surface of the street to the bottom of the foundation for the concrete floor is 11½ feet. The superstructure, with the paving, which is of steel, is very light, but strong, and will not require over 14 inches from the top of the paving to the bottom of the transverse girder.

To permit the running of cars of such a height of body as would accommodate ordinary travel, and at the same time utilize the limited vertical space of the tunnel, the body of the car hangs between trucks of special construction, which admits of the use of a very high wheel to give an easy motion while running at a high rate of speed or in passing around curves. With the open slots located between the surface railroad tracks, both underground and surface cars can be worked from the same circuit as shown in the cut.

CHAPTER IX.

HIGH SPEED INTERURBAN RAILROADS.

During the past year an entirely new branch of electric railroad engineering has arisen, namely, the introduction of electric railroads to replace the present steam roads running from one city to another, as distinguished from local roads in a city. From present appearances this subject promises to be one -of the largest and most important branches in electrical engineering, as it contemplates not only the replacing of the present steam locomotives, but also

promises an increase of speed of more than double
what is accomplished now by steam locomotives,
and to increase the frequency of trains, both of
which will diminish the time it takes to go from
one city to another to such an extent as to have a
most important bearing on the development of
commerce and industries.

The speed of steam locomotives may be said to
have reached almost if not quite the possible maximum, considering both safety and efficiency.
Although isolated cases are recorded in which phenomenal speeds have been made, yet they are
exceptional and not regular, and it is doubtful
whether even with the most improved engines the
speed can be materially increased. The best regular
running time on some roads is from 50 to 60 miles
an hour, which may be considered to be the maximum at present attained in regular service,
although for short distances the speed sometimes
reaches 70 miles. A recent run was reported on the
Bound Brook Line, between Neshaminy Falls and
Langhorn, N. J., of a Wootten steam locomotive, in
which a single mile was made in $39\frac{1}{2}$ seconds, which
is equivalent to 90.45 miles an hour; 10 consecutive
miles were made at 43 seconds per mile, or a speed
of 83.72 miles an hour. Owing to the danger and
inefficiency, however, it is not likely that such
speeds will be attained in regular service on the
present steam roads. The limitations of speed with
the present steam locomotives are due to the heavy
reciprocating parts, which not only are in constant
danger of being strained or broken by such rapid
oscillating motion, but which, as they cannot be

completely balanced, both statically and dynamically, impart to the engine an oscillating motion which is a source of great danger of derailment, and requires an exceptionally heavy roadbed and construction to resist it. But supposing even that the reciprocating parts did not limit the speed, the size and weight of an engine and the amount of fuel which would have to be carried for the enormous powers required at high speed would alone render that system impracticable. In an electric motor, on the other hand, there are absolutely no reciprocating parts and their speed is therefore limited only to that at which a body may be revolved around a shaft; this limit is that at which such bodies would burst, due to centrifugal force. Prof. Wm. Marks, in a lecture on high-speed railroads, states that no matter whether the wheels are large or small, the limiting speed due to bursting admits of a maximum safe speed of 2½ miles a minute, that is, 150 miles an hour. Assuming his calculations to be correct, this is the maximum speed which any car running on revolving wheels can ever attain, no matter what the source of power is.

In actual practice the time it takes to go from one city to another is not dependent only on the speed of the trains, but also on their frequency, as the time lost in waiting for a train might in many cases almost as well be consumed on the road. The present steam system does not adapt itself well for the running of frequent short trains, but an electric system is specially well suited for just that kind of service. If, for instance, short trains or single cars could be run from New York to Philadelphia every ten minutes, instead of a long train every hour, the time it takes to go from the one city to the other would be materially reduced in many cases, even if the present speed was not exceeded.

Much attention has been given this subject of high speed interurban traffic by electrical motors during the past year, and it is not unlikely that in the near future some actual trials will be made. It is reported that a certain large railroad in this country is at present building an experimental line, which is to be run by electric motors and on which experiments are to be made to demonstrate the feasibility of its introduction on a large scale.

The advantages, the possibilities, and the requirements of high speed systems will be best seen from the extracts given below, taken from published opinions. The most interesting literature on the subject which has appeared during the past year is a paper by Mr. O. T. Crosby, read before the American Institute of Electrical Engineers February 24, 1891, a reprint of which will be found in *The Electrical World* of March 14. This paper is a record of some very interesting experiments made on a small scale and includes a project of a road on a large scale worked out in detail, including also a summary of the commercial aspect of such a project. Space does not permit us to reprint it here, but any one interested in the subject should not fail to read it. Attention is also called to the chapter on high speed service in a book recently published, called "The Electric Railway," by Crosby and Bell, in which part of the above paper is included, and is accompanied by much additional information. In

addition to these, an article on train resistance will
be found in a paper on "Limitations of Steam and
Electricity," by O. T. Crosby, read before the American Institute of Electrical Engineers May 21, 1890,
and reprinted in *The Electrical World* May 31, 1890.
In reference to the atmospheric resistance to rapidly
moving trains, a paper was read by the same author
before the West Point Branch of the Military Service Institute, which was published in *Engineering*
(London) May 31 to June 13, 1890, and reprinted
in abstract in *The Electrical World*, May 3, 1890.

Regarding the efficiency of high speed traffic, Mr.
Crosby has made some very interesting calculations
to show at what speed electric traction becomes
feasible. As a result of these deductions he states:
"At a speed of 20 miles an hour and with an efficiency of 90 per cent., and a frequency which is not
excessive as compared with the local traffic, say of
the New York Central Railroad, the electrical propulsion is slightly more economical than steam propulsion. Now, as you increase your speed the relative value of the electrical propulsion increases enor-
mously, and finally at high speed, say 120 miles, the
electrical propulsion is something like six times
more economical than steam propulsion, considering
steam propulsion even possible. But it is not possible to propel ordinary trains by steam at 120 miles
for any length of line worthy of consideration,
because the amount of coal and water that you are
required to carry has reached a quantity that would
require two or three tenders to take care of. You
can see readily that it is a cumulative thing. Every
pound of coal you put in requires some more coal to

be burnt to pull that pound, and you have to carry
it, and you have water, so that when you get to the
high speeds it is out of the question, and we have
nothing to look for at very high speeds, save electrical propulsion. But if we come to comparatively
low speed service, and a comparatively infrequent
service, the steam will beat us."

Regarding heavy freight service, the same writer
concludes that "heavy freight service, unless its
frequency is far beyond anything we have to-day,
can be carried on more economically by steam than
by electricity."

Regarding the point at which high speed electric
roads begin to be practicable, Mr. Crosby gives as
his opinion : " I do not think there is anything whatever in high speed work for say, even 40 miles. I do
not think you can properly do 120 miles an hour for
a run of, say, 30 or 40 miles. It is applicable only to
connecting very large cities. I think you could run
from here to Albany, if you had the New York
Central line to run on; from Albany to Rochester,
from Rochester to Buffalo, and then stopping at the
principal lake cities. It might be found later on
that it would be necessary to make shorter stops;
but the first efforts should be made only with very
long runs."

Regarding automatic systems he says: "I do not
consider it practicable to run a train—that is, of
size enough to make it commercial—without a man
on it. I think the enormous complexity of an automatic system that would take care of the entrance
into a city and the departure from a city, and of all
the varying conditions of service, without a man on

board, is really beyond anything that would stand
up. You might put it all down to work and have
trains on it, but I think the complexity would soon
break it down."

On the same point Mr. F. J. Sprague writes: "I
wish to take the decided position that no automatic
system of high speed service for either passengers,
freight, or mail will ever come to a commercial
issue. I will go further still; the moment you deny
to a man the privilege of traveling 100 or 150 miles
an hour you will have a flood of applications from
people that want the opportunity."

Mr. F. L. Pope, on the other hand, in his report on
the Portelectric system (which will be found described below under the heading of miscellaneous
systems) states: "The opinion has recently been
publicly expressed by an electrical engineer of reputation and experience that no automatic system of
high-speed service for either passengers, freight, or
mail, can ever come to a commercial issue. No
arguments were adduced in support of this dictum,
and, for my part, I am wholly unable to see any
good reason why this assertion should have been
made in such an unqualified and authoritative manner. The experiments conducted under my direction at Dorchester are, in my opinion, sufficient to
prove the contrary."

Regarding derailment at such high speeds, a
prominent civil engineer, at one time professionally
connected with the Lake Shore road, expressed the
opinion that safety from derailment at very high
speeds would be best secured by very slightly cur-
ving the line of the road just sufficient to cause the

flanges of the wheels to bear constantly against **one** side. With that construction he should expect that any possible speed up to, say, 200 miles an hour, that could be got from the motors, would be perfectly safe. He was led to this conclusion from long observation of the performances on the Lake Shore road of railroad wheels at high speed, on which road some of the best running in this country has been done.

Regarding the traction co-efficient, Dr. P. H. Dudley, referring to some experiments made by him on the resistance of different trains on steam roads, states: "The experiments were made several years ago, and not on as good tracks as we have at the present time. The resistance of the passenger train is given by the formulæ in books as 17 or 18 pounds per ton. I found on the Lake Shore & Michigan Southern Railway and the New York Central & Hudson River Railroad that, with trains of about 250 tons, the resistance was only from 10 to 12 pounds per ton at speeds of from 50 to 60 miles per hour. The resistance per ton is not nearly so great on those long trains as on the short ones. With trains of two and three cars, sometimes it ran as high as 35 or 40 pounds per ton, but with long trains it ran down to about 10 and 12 pounds per ton.

"The greatest variable that we find from day to day over the same track and same train is the wind resistance. I found in one special experiment I made in 1878, to see why the New York Central was not able to make its time with its trains, that on a still day, with the schedules they had, they were able to make time, but a wind of 10 or 12 miles would retard the trains, so that in running from Buffalo to New York they would lose two hours. The trouble at that time was that the boilers were not large enough to quickly generate the steam required to meet the increased resistance. The trains would run up to a speed just about the capacity of the boiler, and that would limit the speed they could maintain. Therefore, when a head or side wind was blowing against them the speed of the train would be reduced very materially.

"The fastest speed usually found on trains is in the local passenger service, for 5 to 20 seconds— especially those stopping every two or three miles. They often attain a speed for a few seconds as high as 55 or 60 miles per hour. Then they shut off steam and apply the brakes, to stop the train. The heavy trains of eight to ten cars, running long distances, rarely attain 60 miles an hour—that is, on the ordinary schedules.

"The greatest improvement that has been made in decreasing train resistance is in the improvement of the track by bringing up the standard of excellence and adoption of heavier and stiffer rails. I have been making a number of heavy rail sections for different railroads, and instead of making them simply heavy I have made them very stiff, which has reduced the deflection or wave motion under each of the wheels. Comparing the resistance of the Chicago Limited Express on stiff 80-pound rails with that of 65-pound rails which have been on the track some time, it makes a difference of 75 to 100 horse-power per mile. I have just made some 95-pound rails for a road which will be very much stiffer than the 80-pound rails. We find now that the condition in which they are keeping the line is so perfect that there is scarcely any oscillation on the best roads, and there is very little difference in the oscillation now in riding on a tangent or a curve. I should prefer a tangent for smooth riding, although the wheels press against the rails on curves. Unless they are properly elevated, there will be some oscillation, though our present trains move very steadily on the best tracks. I have made careful experiments, and with a track in good condition the heavy cars ride very steadily, leaving little to be desired."

Regarding the difference of the traction co-efficient of high speed roads and the ordinary street roads, Mr. Sprague states: "If you are dealing with street car service, it is one thing. If you are dealing with high speed, clean rail service, it is an entirely different thing. I have found, so far as my practical experience goes, that the dirt on the track, the snow and mud, and the grades, were the questions that determined the power of the motors. When you consider that with 20 pounds traction per ton, the work on a one per cent. grade is equal to the work of traction, and that our street car service demands 12, 13, and even 14 per cent. grade work, the question of rail traction on a level amounts to very little in street car motor design, but it does amount to the greatest possible importance in high speed service."

In discussing the subject of rails, which is of such vital importance in high speed work, Dr. P. H. Dudley gave some interesting figures and results ob-

tained by him from a series of elaborate experiments.
In describing the results obtained from improved
forms of rails, he says: "About the best results are a
reduction of nearly 100 horse-power per mile compared with the ordinary 65-pound rail.   I have
designed some 105-pound rails which are nearly
1,100 per cent. stiffer than the 80-pound rails.   We
will probably save on the fast express trains nearly
200 horse-power per mile, as compared with worn 60
or 65-pound rails.   We have not rolled such a large
rail yet; but it will not be long, probably, before
that will be done.   All trials show a very material
reduction in train resistance with the heavy rails,
as we increase the stiffness.   I think it will be perfectly safe on a track of the 105-pound rails to run
120 miles an hour.   I have ridden several times 75
miles an hour on heavy rails, and did not feel it
nearly so much as on the light rails when running
45 miles an hour.   There is scarcely any oscillation.
The track is in perfect line and the joints are all
kept up.   One serious trouble with many of the railroad people is that, while they want a heavy rail,
they never consider the question of stiffness.   I distribute my metal so as to make the section very
stiff, not allowing so much in depth of the head of
the rail as usually is done, but widening the head,
because when these rails have been in the track a
few years they become worn very unevenly, and
must be removed from the track.   I am making the
heads on that account very much broader and the
sections stiff, so the deflection of my 75 and 80-pound
rail is less than one-tenth of an inch in the track.
We mark all those rails for any deflection exceed-

ing three thirty-seconds of an inch, under a load of six tons per wheel. That is about as close as the trackmen can work on a 65-pound rail. On an 80-pound rail they can work closer, but require great skill." He described the condition in which tracks were found, after continued service, for illustrations of which see *The Electrical World*, p. 230, March 21, 1891. To study the unevennesses in tracks he uses an ingenious device which is carried on the moving train and which ejects paint on the rails wherever there is a certain deflection.

Regarding the braking of trains Mr. E. P. Thompson states: "I found in some tests made by me in behalf of a brake company upon a railway in Pennsylvania that if it needs a certain amount of energy to start a train it needs about nine-tenths as much to make an emergency stop."

Referring to the possible recovery of energy in braking a train Mr. F. J. Sprague states: "From careful calculations I know it to be possible to return about 40 per cent. of the total energy used back to the line, and the facility with which this can be done increases very rapidly with the size of machine which is used. Where you can use shunt machines, and excite your field magnets up to a very high saturation, and have a very reasonable latitude of variation, the criticism that was made that there would be a loss in energizing the field magnets more than when in using series machines I do not think holds particularly good, for this reason, that the total energy used in the field magnets forms a very small proportion of the total amount of energy used on the line, and if the return

energy is 40 per cent. of the total, and the field
energy of the magnet is only two or three per cent.
of the total, the value of this method seems apparent."

Descriptive.

One of the most complete projects published, on
rapid transit between cities, is contained in a very
interesting lecture delivered by Mr. Carl Ziper-
nowsky, at the Frankfort Congress, on a proposed
road between Vienna and Budapest, entitled "Electric Railroads for Rapid Transit Between Cities." It
contains so many points of interest that we give a
translation here in full, as follows:

Introductory.—The need of some means for increasing the speed of trains between cities, above
that which can be accomplished by the present railroads, is becoming a question of such importance
that it is now only a matter of a short time before
something will be done, and it is for this reason that
we have occupied ourselves with a proposed electrical railroad to accomplish this object.

All persons who have much traveling to do will
appreciate the great loss of time while on the road.
To increase the speed of the present system of railroads above 100 kilometres (62 miles) an hour may
be excluded as impracticable. The reciprocating
motion of heavy parts of express locomotives gives
rise to oscillations which greatly strain the rolling
stock as well as the roadbed. For this reason the
present system of steam locomotives is limited as
to maximum speed. With electrical motors there
are no reciprocating parts, and the limitation of the
speed due to the reciprocating motions does not

exist; the speed can, therefore, be increased above the present limit, without increasing the strain on the rolling stock or the roadbed. The electrical locomotive needs no coal, no water, no tender, no generator of power, only a motor whose rotating parts are directly on the axles.

The electrical railroad has these advantages: it offers the possibilities of a new system of transit which, owing to the great increase of speed which it admits, adapts itself particularly well to the rapidly inreasing demands for rapid transit. As soon as the possibility is shown, the working out of the details will soon follow, as the very rapidly increasing demands of commerce make it necessary that everything possible be done to accomplish this end, and it is certain that within a few years such railroads will be built between some large cities, which are closely connected commercially. It may also be predicted that whole continents will be crossed by such railroads.

We chose for our project the two chief cities of the Austrian-Hungarian empire, Vienna and Budapest, not only because there is a very large intercourse between them, but also because this lies on the throughfare between the centres of the oriental and the western countries. An electrical rapid transit road between Budapest and Vienna will form the beginning of a through line from the east to Paris and Havre on the one hand, and to Berlin and Hamburg on the other.

This projected road is furthermore particularly well adapted for an experimental road, as it introduces all the difficulties encountered in building such a road through different kinds of country. The crossing of the wooded hills of Bakonyer, the very low islands of a branch of the Danube, and the hilly country between the Danube and the Leitha Mountains, gave us the opportunity to show how we would build such a road over flat lands, hills, and through mountains, and what fundamental principles we would lay down.

Before going into this, we wish to mention that, from the nature of such a system, there would not be trains of cars at long intervals, but on the contrary single cars for few persons, following each other in quick succession. This is preferred not only in order to have frequent connections between the cities rather than occasional fast trains carrying many people, but more particularly for technical reasons, namely, to have the greatest economy in the power used per car, and to equalize as much as possible the consumption of power over the whole line. The smaller the weight of a train, the smaller will be the motors, the smaller the current consumed per train, the simpler, surer, and cheaper will be the leads. The smaller the units, the surer and the more profitable will be the working of the system, because one can prepare for the variations in the traffic during the day, and can thereby always obtain a more favorable relation between the dead and the live loads. Small units at short intervals have the additional advantage that the leads are more equally loaded than they would be with long heavy trains. The utilization of the leads and the machines becomes more nearly constant. and the whole plant will be more efficient. The increase in the number of employés necessitated by such a system is negligently small as compared with the very greatly increased cost which would be necessitated by running the plant under unfavorable circumstances. These are the reasons which have led us to adopt this tramway system in preference to the usual railway system.

In order to design the cars and roadbed, it was necessary first to determine upon the speed, the shortest time intervals between the trains, the number of passengers and the nature of the cars, and to base on these values the size and shape of the cars, the power required and the size of the motors, the total weight of a car, the roadbed, the method of distributing the current, etc.

As to the speed, we wish to reach the maximum which appears to be possible to attain with simple adhesion and a modified roadbed. The speed can be increased only up to the limit of the strength of the materials, especially of that of the wheels. This limit is shown by calculation to be about 250 kilometres per hour (155 miles per hour, or about 2½ miles per minute). The circumferential speed of the wheels of 2.5 metres (about 8 feet 3 inches) diameter, becomes so great, being nearly 70 metres (229 feet) per second, that the bursting due to centrifugal force comes into consideration. It may be shown that even with wheels made of discs it is not safe to exceed the circumferential speed of 70 metres per second, as even the best materials at our disposal do not allow a sufficiently large factor of safety against rupture.

The adhesion also limits the speed to which we

may go in order to be able to propel a car with sufficient reliability. But this limit cannot be readily determined, as the grades, nature of the materials, weather, and similar factors vary very greatly. It is not possible to determine, therefore, the speed at which propulsion is still possible, and we are compelled in this case to make an approximate estimate on the basis of the greatest grades.

We will assume that the maximum speed attainable with single cars is 250 kilometres on the level and 200 on the grades. The intervals between the cars will be regulated by the traffic and the size of the car. Here, too, there is a limit due to safety; no car should follow another at a less distance than that at which it can be brought to a stop if something should delay the one ahead. The stopping and signaling evidently depend on the facilities provided for this purpose. If we were to employ for such rapid transit the same system of signaling as on the present roads at 100 kilometres speed, we would run the risk of their failing at this high speed, at which both seeing and hearing are rendered difficult and unreliable. It is, therefore, necessary to adopt an entirely new system of signaling specially designed for such speed. Furthermore, there must be provisions for the case that the signals should fail to be seen from the car, in which case the signaling station must be able to stop any car even without any further signaling, which is easily accomplished on electrical roads by simply introducing switches controlling the main current. The cars may be said to be controlled from the outside as in the case of pneumatic tube projectiles.

The minimum interval between cars, therefore, is determined by the condition that a car cannot move any farther, by virtue of its momentum and a possible grade, after the current has been turned off, than a certain minimum distance, which must be kept, for safety's sake, between it and the preceding car, even when at rest. This distance can be made less the better the braking facilities on the car, and the surer the engineer is to notice when the current has been cut off by the signaling station. It is evident, therefore, that the smallest interval between cars can be determined only in connection with the braking facilities, but it should not be made less than 10 minutes. This minimum interval is, of course, only for those hours of the day when there is most travel.

The capacity of the cars must be such as the densest traffic demands; but on the other hand the cars should not be any larger or heavier than is necessary, for reasons already given, for which reasons it would also be inadvisable to couple several cars into a train. The determination of the dimensions of the cars is difficult to make until one has had experience in how far the public would patronize such rapid transit. But we believe that it is safe to estimate 40 seats to a car, and intervals of from 10 to 60 minutes. During the times of greatest travel 200 persons could be carried per hour, which surely is considerable; during the times of least travel this would be reduced to about 50 per hour.

Besides the transporting of passengers, such a system would be of great use for the mails. The transportation of freight other than the small baggage of the passegers can hardly be considered, as the cost per pound would ncessarily be too great for any but very exceptional cases; even for heavy baggage it would hardly pay.

The two generating stations are intended to be in the cities Banhid and Zurndorf, which are about 60 kilometres from Budapest and Vienna respectively. From there currents of 10,000 volts are to be sent along the whole line on poles, from which wires branch off to secondary stations all along the roads, which are at the same time signaling stations. In these secondary stations this high tension alternating current is transformed into one of lower tension or by means of one of our transformers into a continuous current. It will depend entirely on the results of some experiments with the motors for the cars which of these two currents will have the preference.

Construction of the Car.—The car was constructed in the car factory of Ganz & Co., of Budapest. It contains 40 seats for passengers, two toilet rooms, and some available spaces for the mail. It is 45 metres (148 feet) long, 2.15 metres (7 feet) wide, and 2.20 metres (7¼ feet) high; the two ends are shaped somewhat like a parabola, in order to reduce to the smallest possible amount the air resistance, which at such speeds absorbs the greatest part of the power.

The two ends of the cars are exclusively for the motors, and are inaccessible to passengers. They are separated from the main part by partitions with doors and windows. They must also be separated from the engineer's stand, on account of the very violent currents of air which are generated in this space. Each of the motor rooms represents a separate truck.

The framework of the car is a system of beams which rest on the two motor trucks, and in the body of the car form a sort of truss bridge, with cross braces at every 1.5 metres. The bridge construction adapts itself best to the conditions. The body of the car rests on the two pivoted trucks, and is supported by 16 pairs of evolute springs, which are held in cast steel telescopic cases. The support is such that both trucks can turn, corresponding to a curve of 1,000 metres radius; the side motion of these cases inside of the rails is in that case 6 millimetres.

Each truck has two axles, and each axle has its motor, making therefore four motors to a car. The magnets are secured to the frame of the truck and the armatures are directly on the axles. The driving wheels (all of the four pairs of wheels of a car are drivers) were made as large as possible, and have two flanges, the outside one being merely for safety sake against derailing, and is 5 millimetres distant from the edge of the rail. The inner flanges also have 5 millimetres' clearance, to allow for the heating and the expansion of the axles. The wheels are made of two solid conical discs, secured on the outside edge to the tires in such a way that the tire is readily replaceable. On account of the great strain on the tires they must be very carefully made. The bearings are of special importance; the pressure on the wheel being nearly 7,500 kilogrammes, and the speed 600 revolutions, an entirely new construction was chosen. The bearing boxes are solid, notwithstanding the attending disadvantages. In the exact middle of each truck there are two contact wheels, which run on separate rails, carrying the current. These wheels are flanged so as to be guided by the rail; it is very important that they be exactly in the centre and that their axial motion be as little as possible. As the currents are necessarily quite great, the nature of these contacts is a very important matter. The contact wheels must have a large diameter in order not to have too great a speed of revolution which would introduce difficulties at the bearings; they must furthermore be light, so as to follow any motions readily, and must rest on the rails with some pressure. We have made them similarly to the driving wheels, of two steel discs secured together, holding a readily replaceable tire of bronze. The three bearings of each of these contact wheels are secured to the truck by means of three pivoted arms and pressed on to the contact rail by means of three vertical springs. The current is taken off of these wheels by means of massive copper blocks which rest against the wheels. Their shafts are made of two parts, and insulated in such a way that neither the shaft nor the bearings carry current. The contact wheels and electric motors are accessible from a narrow bridge which runs the whole length of the motor room above the axles.

The braking mechanism must be particularly well designed, as it is of vital importance for safety's sake, with a speed of 250 kilometres an hour, to absorb or destroy a momentum representing 60 tons in the shortest possible time, and thereby to stop the car. The air friction itself acts as a powerful brake, being equal to about 200 horse-power. On the level all other resistances are very small as compared to that of the air. It is evident, therefore, that the first braking action on stopping the current will be that of the air. But this applies only to the first moments, as its action diminishes very rapidly as the speed becomes less. A second powerful braking action may be produced by connecting the terminals of the motor to a resistance, which may be put under the bottom of the car, and to let the motor act as a generator. It is immaterial, as far as the necessary switching arrangements are concerned, whether continuous or alternating currents are used, and no difficulties present themselves. As the speed diminishes to about a half, the motors which were originally in parallel should be capable of being connected two in series and two in parallel, and at, say, one-quarter speed all four in series.

This braking action, although effective, is not suficiently powerful for speeds below 30 kilometres per hour (about 18 miles). There must be an additional mechanical brake, for which we prefer the usual Westinghouse air brake, as the safest and quickest. For this purpose there are two air chambers and eight cylinders on the car. The dimensions, capacity, and pressure from these are quite ample, and the loss of pressure can be replaced either by hand or by means of a small electric motor. The two reservoirs are made of 4-inch pipes, each 30 metres (96 feet) long.

The car is equipped with air buffers, the heat generated by their action being used to generate steam, which is then allowed to escape. The cars furthermore have couplings, so that a disabled car could be drawn by another. The headlight for the night must be so arranged that it can be directed. The illumination of this must be so powerful that any obstacle, as for instance a fallen tree, can be seen at a sufficient distance to enable the car to be stopped. This distance will be about 2 kilometres (1¼ miles); the headlight reflectors will, therefore. have to enable one to see even in bad weather at this distance.

The illumination of the car is to be effected by means of incandescent lamps supplied from the main current, through a special apparatus which equalizes the variable potential. The heating is by means of two coal stoves, which may be placed between the two walls. The windows have double panes of glass, and are arranged so that they cannot be opened. Ventilation is effected by suitable pipes on the roof of the car.

Construction of the Road.—In the above we have endeavored to give the general principles of our projected system which form the basis on which to work, namely, the mode and the means for a system of rapid transit. We will now consider the conditions which are thereby necessitated concerning the construction of the road itself.

As already explained, the grades at such a speed of 200 kilometres do not offer as many difficulties as the curves. In order to mount grades with greater speed requires merely a greater power, and it can always be so arranged that this increase of power can be obtained at the proper places. The speed is limited on the curves, for which reason only curves of large radius can be used. For that part of the road over which cars are to be run at full speed we have limited ourselves to curves of not less than 3,000 meters radius (9,840) feet; where the nature of the land does not permit this the speed will have to be diminished. The side pressure on the curves is to be avoided entirely by raising the outside rail. By this we mean, of course, that the difference in height of the two rails is obtained partly (one-half) by a lowering of the inside rail and partly (one-half) by raising the outside rail. The difference in the heights of the two rails for a speed of 200 kilometres and a curve of 3,000 metres radius is 148 millimetres (5¾ inches), at which the resultant of the centrifugal force and the weight is normal to the rail. The greater this difference, the more difficult will be the passage from a straight stretch to a curve. It will not be possible to make this difference greater than 180 millimetres (7⅛ inches) without endangering the safety of the passengers at the curves. In laying out the road the curves and the greatest possible speed will have to be considered for each case, taking into account the grade, direction, etc., and the rise of the outer rail determined from this.

The great attention which has to be given to the curves, in laying out the road to comply with the contour of the land, increases the difficulties greatly, and for this reason one should not be too particular to avoid grades. In our projected road from Vienna to Budapest, we have not hesitated to use grades of 10 per cent., and we assume that the speed on these grades shall not be less than 200 kilometres

(124 miles); the power, however, will be twice as great as for the same speed on the level. Steeper grades than 10 per cent. would increase the weights of the more powerful motors to an unfavorable degree, and they are, therefore, to be avoided if possible; if they cannot be avoided the speed on these grades will simply have to be reduced.

It will be seen from the above that the average speed in hilly and mountainous districts will not be more than 200 kilometres (124 miles) per hour; on down grades and on the level the time lost on the curves and up grades is made up by a maximum speed of 250 kilometres (155 miles). .

Having now considered the laying out of the road we next take up the construction of the road itself, which is so intimately connected with that of the cars. We are, above all, safe against derailing, on account of the large diameter of the drivers and the nature of the construction of the cars, whose pivoted trucks can be guided so much better (having but five millimetres or three-sixteenth inch play) than is possible with the usual rigid shafts.* Furthermore, the great length of the car is the best means for preventing the lateral motions of the car, and thereby avoiding the chief cause of derailing. A second safeguard is secured in the second flange on the drivers, both of which are furthermore made 50 millimetres (two inches) high. Finally, to guide the car in case the wheels should leave the rails, we have provided that the framework of the car extend down outside of and below the top of the rail.

---

* The usual railroad cars in Europe have no pivoted trucks, being built more like the American horse cars.—ED.

Besides this, the body of the car is carried so low that the distance to the current-carrying rails is only 100 millimetres (four inches), and therefore in case of derailing these rails would still act as a sort of guide and support for the car.

But the best protection against derailing is the particular roadbed and rails which we have chosen. The latter is a 180-millimetre (7⅛ inch) high Vignol rail, weighing 50 kilogrammes per metre (about 33⅓ pounds per foot), which is screwed down to cast-steel sleepers by cast-steel supports on both sides of the rail. The latter have their surfaces planed and grooved so as to assure the proper width of gauge beyond question. They are placed at distances of 1 metre (3¼) feet apart, and are bolted down to a continuous foundation of concrete. The rails are supported along their whole length on the foundation, so that in case of a rail breaking it is still held in place. The current rails are constructed as a sort of air line, supported on insulators about 500 millimetres (20 inches) from the ground, by means of cast-iron carriers held in supports cast for them in the steel sleepers. The whole roadbed must be constructed with the utmost care and precision even in the minutest details, and the rails must be screwed to the foundation. Such a construction, on a continuous foundation, increases the cost of the road very greatly, but we consider this as unavoidable, on account of the great necessity to guard against all sources of danger. It is above all the weight and stability of the roadbed which insures safety. It is out of the question to make the bed of loose stones, as this is altogether too elastic. Almost all the cases of derailment can be traced back to changes in the bedding; most of the accidents which have occurred can be traced to unequal settling of the roadbed, and similar causes. In our system there is furthermore another reason for having a very heavy, massive roadbed; the shock or blow which such a heavy car exerts at that great speed must be met by some correspondingly heavy and massive body, and only then will the roadbed stand the strains without danger. The car must roll perfectly quietly on its absolutely solid foundations.

The desired rigidity is secured only when the rails and sleepers are actually bolted to the foundation. This foundation may be made of two continuous underground walls of masonry; where the road has been filled in, these walls must be made much stronger than in cuts or where they are built in the natural ground. For that reason there should be no high embankments, as they keep on settling even after several years, and therefore do not offer a sufficiently secure foundation. For bridges the conditions of the road are more favorable than for steam roads, as it is necessary to consider only the maximum weight of two cars passing each other, say about 120 tons; the rigidity must, however, be considered on account of the violent action, almost like a blow, which such a rapidly moving car exerts. Stone viaducts will be found to be necessary much more frequently than for steam railroads, partly because, as we have mentioned, high embankments cannot be relied upon as sufficiently safe; partly, also, because the former will be cheaper than the latter. The reasons for this are, that for such a system of single cars instead of trains, long roads must unquestionably be built with double tracks, as turn-outs are out of the question, partly because of the loss of time, which is just that which our system is intended to avoid at all costs, and partly because switches would have to be traversed at great speeds, which would heighten the source of danger very greatly. Anything but a double track road is therefore entirely out of the question, except on very short roads on which there is only one car at a time. The two tracks must be at least 10 metres (33 feet) apart, on account of the intense air currents and air friction, the shock from which might become a great source of danger at a smaller distance than this. With a distance between the rails of 10 metres, a high embankment would become so very expensive that its cost would be greater than that of two independent parallel viaducts 10 metres apart. We proposed for the latter to use the best construction of cement and iron on piers with spans of 12–15 metres (40–50 feet), and a width of 2½ metres (8¼ feet). We have found from calculations that for the line from Vienna to Budapest such viaducts are not more expensive than embankments when the height of the track above the natural surface of the ground is 6 metres (19¾ feet), even when the wide space between the two tracks is not filled in.

The road, of course, must be made inaccessible to all pedestrians except the railroad employés. Grade crossings are out of the question. To save the trouble of clearing away the snow, we propose to elevate the top of the foundation of the roadbed 500 millimetres above the lower level of the road, in order that the snow may be partly, at least, blown away by the intense current of air accompanying each car.

As already mentioned, great attention must be given to the signaling. As it may be necessary that the signaling station has to stop a car even when the locomotive engineer has failed to see the signal, all signaling arrangements must be so designed that with each signal the current supplied to the rails must be correspondingly turned off or on by the signaling station.

To accomplish this we lay down the following conditions:

1. Along the whole line signaling stations must be erected, which for a double track road must not be at a greater distance apart than two kilometres (1¼ miles).

2. The rails carrying the currents for each section must be cut and insulated from those of the next section; the leads supplying these sections at these places must pass through a controlling and regulating device.

3. This device must show when the current is greater than the normal, by reason of two successive cars being closer together than the normal distance, and it must furthermore diminish the current for the second car, so as to compel it to go slower until the distance between it and the one ahead is again normal.

4. Every signaling station must be capable of sending its signals to the two neighboring stations.

5. Every signal should be such that the engineer can readily see it.

6. The signals must be quite long, in order that he may see the colors or lights as stripes or bands.

We propose to signal with three bands, having the following significance:*

Three bands, "stop;" two bands, speed of 50 kilometres; one band, speed of 100 kilometres; no bands, full speed.

7. Besides the above there should be signals for covering stations, and a telephone connection between the stations.

Power Required.—The power required for propelling a car of about 60 tons, and having a cross section of nearly 5 square metres (about 50 square feet), is very considerable, and it is evident that the greater part of the power is required to overcome the resistance of the air. In the absence of reliable empirical data regarding the various resistances encountered by a railroad train, we have taken up each of them in series, as far as concerns the scope and object of the present project, and give the results here as follows:

The air resistance was determined lately by Crosby (see *The Electrical World*, 1890, vol. XV., p. 346) to obtain a safe basis for calculating the projected rapid transit road from New York to Chicago. He obtained figures for the moving of differently shaped bodies in air, which are considerably smaller than those of the formulæ heretofore used.

As Crosby's experiments were made with great care, we can unhesitatingly use his empirical for-

---

* We would suggest that it would be more rational to reverse this set of signals so that if the engineer fail to see the signal, or if one or more fail to act, it would cause him to stop or diminish his speed rather than to increase it, as he would with the above arrangement.—ED.

mulæ for the resistance of air at the face of the car, namely:

$$P = 0.1441 \ V$$

in which P is in pounds per square foot· and V the velocity in miles; and

$$P = \frac{P + \cos. \ \theta}{2}$$

for the pressure against surfaces moved at an angle.

According to these formulæ one can assume that the air resistance of a well-constructed car for 200 kilometres, average speed will not exceed 250 horse-power. If steep grades (10 per cent.) are to be ascended at this speed, it will require for a car weighing 60 tons nearly 450 horse-power.

To this must be added the resistance of curves, the air resistance on the lateral surfaces, the rolling friction, bearing friction, loss of power by lateral oscillations, etc., which, however, cannot be calculated for want of experimental data, but which are no doubt quite small as compared to the others, and may, therefore, be safely estimated at 100 horse-power.

The maximum power required for a car is according to this 800 horse-power, and every car would, therefore, have four motors of 200 effective horse-power each. Each car will, therefore, require in favorable weather about 260,000 watts when the road is level, and up to 600,000 watts in ascending grades.

The voltage of the working current could not be chosen higher than 1,000 volts, as interruptions and complications are too apt to be caused by poor insulation, etc., at higher voltages than this. Besides, all parts of the circuit on the car should be capable of being handled without danger to life.

The amount to be transmitted to the moving car must therefore be from 260-600 ampères, which requires very good contact surfaces.

We may assume that the above problem may be considered solved, as far as the railroad engineering is concerned. There remains to be considered what the best means are for the electrical transmission of the power.

We reserve this question, however, for our reconsideration in view of the results which are to be obtained from some experiments which we are at present making with a new form of motor specially designed for railroad purposes.

To the above paper the following additional facts taken from other sources may be added: The gauge of this proposed road is to be 1.45 metres (4 feet 9½ inches) the same as that used in Europe at present. The cost of such road is estimated to be about two and a half times that of an ordinary railroad. The car is shaped somewhat like a cigar pointed at both ends. The seats are like those in the usual horse cars; the entrances are at side doors where the car proper joins the motor rooms. The proposed speed is about double that of a train at present running regularly between Brighton and London, on a 6-foot gauge. The cost of this road is estimated at from $16,000,000 to $20,000,000. The total distance is 250 kilometres (156 miles). On the present steam road it takes an express train four and a half hours. In connection

with this project Prof. Elihu Thomson is stated to
have given the opinion that the highest attainable
speed might reach 300 kilometers (186 miles) an
hour, or 3.1 miles a minute.—ED.

CHAPTER X.

MISCELLANEOUS SYSTEMS.

Patton Self-Contained System.—Quite a departure
from the usual plan is made in this system, as it
does away with trolleys, conduits, power stations,
etc. It consists briefly of a car equipped with a
small gas engine, a dynamo, accumulators, and a
motor. The source of power is carried in the form
of gasoline. Though it may seem cumbersome, it
nevertheless is interesting as a radical departure,
and although it hardly looks as if it would replace
the more usual systems, it nevertheless has a number of points in its favor.

The experimental car which has been in daily service at Pullman, Ill., is said to have attained a speed
of 11 miles an hour while drawing a loaded trailer.
The apparatus used consists of a vertical 10 horse-
power gasoline engine, a 10 horse-power compound
wound generator driven by friction, a battery of
100 accumulators and a 15 horse-power single reduction motor suspended from the front axle. The cells
are placed under the seats that extend across the
car, and are seldom if ever removed from their position. The gas engine and generator occupy a space
the width of the car by 44 inches in length, rela-

tively the seating capacity of eight passengers where two seats extend across the car. The base of the engine is supported on an iron frame placed six inches above the rails; the muffler extends lengthwise of the car. The engine and generator weigh about 1,500 pounds, and the section of the car containing them is inclosed to the ceiling. The gasoline is stored in a tank fastened to the ceiling and is fed by gravity to the burner in the engine through flexible rubber tubes, the rate of consumption being one and a half gallons an hour, irrespective of the load, the fluid costing six cents a gallon. Thus, to run continuously for 18 hours would require a tank capacity of 27 gallons, or a tank holding 14 gallons to be filled twice a day. As the gasoline is automatically fed to the engine no special attention is required, and only a car operator and a conductor are employed, as the engine continues in operation throughout the day, whether the car is in motion or not, and is only stopped when the car is withdrawn from service.

When the engine is in operation the current generated by the dynamo passes to the accumulators if the car is standing, but when the car is in motion the current passes to the motor, both generator and battery being placed in multiple with the motor. The movement of a simple switch handle controls a pole changer, and also cuts in or out the necessary resistance employed. On a level track the motor is supplied with current only from the generator, but on rounding a sharp curve or ascending a heavy grade, where the potential of the generator falls to that of the cells, the latter are automatically placed in circuit and assist in supplying the necessary quantity of current to enable the motor to perform its work. Thus on an ordinary track the cells would only be in service about one-sixth of the time the car was on duty, and there would always be a sufficient supply of current in the cells to run the car for two hours or more independent of engine and generator.

The field in which this motor car can compete is in the service in the small towns, where the cars run "at will;" in the smaller cities, where short competing lines practically prohibit the outlay for cable systems or electric power house and overhead equipments; and in the larger cities, where the trolley will not be tolerated.

A Novel Electric Railway for the World's Fair.— This novel railway will differ from ordinary railways, in that the passengers are transported on a movable sidewalk instead of by cars of the ordinary type. This sidewalk is to be constructed on an elevated structure 25 feet high and 900 feet long, in the form of an ellipse, and is to consist of 75 cars, each 12 feet long, connected together, making one solid train. There are to be constructed two parallel sidewalks, one running at the rate of two miles an hour, the other at four, both walks moving in the same direction. The passengers can step from the stationary walk to the one which moves at the rate of two miles an hour, and if it is desired to move at a greater speed they can step from this walk to the one running at four miles an hour. The passengers can safely walk upon either of the movable sidewalks while in motion if desired. The carrying capacity is intended to be 30,000 passengers per hour. Three of the 75 cars are to be equipped with two 15 horsepower Thomson-Houston railway motors, each mounted upon trucks with wheels 18 inches in diameter. As the car platform, or sidewalk, is arranged it is perfectly level with the stationary walk, allowing the trolley wire to be placed beneath the surface of the platform, and the current taken therefrom by means of small trolleys attached beneath the car floors. The operation of this train of cars will be arranged in a novel manner, doing away with the use of motor men, the entire train being controlled and operated by one man. There will be constructed at a central point, at one side of the track, a controlling station, which will contain a main switch, reversing switch, automatic circuit breaker, lightning arrester, ampère meter, and rheostats, all arranged so that they can be operated by the attendant from that point, who will have the train under perfect control.

Parcel Exchange System.—A London correspondent gives the following description of a parcel exchange system devised by Mr. A. R. Bennett, of London. In many of the large towns of England, the vehicular traffic is so heavy that in order to avoid absolute blocking of the thoroughfares the collection or delivery of goods is forbidden in cer·tain localities during business hours. The results of this restriction are that trade suffers, and warehouses have to be made of larger capacity than would be necessary if the free receipt and dispatch of goods were permitted. With a view to overcome this difficulty, and to allow of comparatively small packages being handled at all times, this scheme was worked out, founded upon the "telephone exchange" principle, by which parcels could be readily interchanged between any number of buildings, no matter how widely far apart they may be situated. The following is a brief résumé:

It is proposed to effect this system of interchanging by the establishment of a number of miniature underground electric railways, radiating from a central station, and having branch lines or sidings into all the buildings to be served. According to this plan, the railways would be laid in tubes of a rectangular section, and would be so arranged that the down track would occupy the lower portion, and the up track the upper portion of the tube. The tubes would be made sufficiently large—say two feet wide by three feet high—to allow a man to creep through for examination and repairs, and in order to afford space for this the rails would be laid, not on cross sleepers or ties, but on brackets fastened to the walls of the tube. Trucks actuated by electromotors would run on the rails, the current being obtained either from one of the latter or from a separate conductor laid parallel with the track. On the down journey the current would be collected by a kind of shoe pressing against the under side of this conductor, and on the up journey by a second shoe or collector. Separate shoes are, however, provided and connected with the motor so that a truck could not travel in the wrong direction. The size of tube suggested would permit of trucks 20 inches wide by 14 inches deep being used, and their length might be considerable, but it would be regulated by the radius of the curves. Each train would consist of a motor truck, and one, or perhaps two or three, trailers or other trucks. The generating and operating station would be established in a suitable locality; in a large town there might be several. The station would contain the engines, boilers, and dynamos, and might also be used as an electric light station. Here would be arranged various turntables for the interchange of trains between the tubes, while sidings would be provided for empty trucks.

Regarding the delivery of goods, connection with the premises of subscribers would be made by short spurs or sidings diverging from the nearest main tube. At the junction of the branches with the main track, switches similar to ordinary railway switches would be placed and controlled by means of electro-magnets by the operator at the central station. Various methods for finding and working any switch with certainty and rapidity are proposed, and also for ascertaining that the switch has been put over or *vice versa*. The sidings into subscribers' buildings would consist of down and up tracks, but where space is available they would be caused to diverge after entering the building and ultimately meet on one track, so that trains might be shifted from the down to the up track without lifting them off the rails. Various arrangements are provided for signaling and for informing the operator or operators of the progress or position of the trains, and for the return of loaded or empty trucks from subscribers' sidings or on the main up line. The starting levers could be interlocked with the levers controlling the siding switches, so that a following train could not leave until the switch for the preceding one had been restored to its normal position.

The inventor claims that the details do not comprise any device which has not been thoroughly tested in the telegraph and signaling department of the post office and railway companies, or in connection with electric traction. He is of opinion that the system of electrical parcel exchange as proposed would be invaluable to the various post-offices, and to parcel receiving and great dispatchers of small packages; buyers could, he says, telephone for samples, hotels and restaurants could telephone for and receive in a few minutes supplies they may be short of, etc. As a parting shot, the inventor gives a friend's suggestion that a mother could send her baby bodily to the doctor via the central station, and receive it back with "a bottle of medicine in its fist and a mustard leaf on its chest."

The Electrical Transportation System of the New England Portelectric Company.—This system consists in general of a small projectile-like car operated entirely by electrical means from stationary stations. A very good description, too long to be reproduced here, in the form of a full report by Franklin L. Pope, will be found in *The Electrical World*, May 23, 1891, p. 375. Other descriptions, with illustrations, will be found in the same journal for May 4, 1889, and October 18, 1890.

CHAPTER XI.

GENERATORS, MOTORS AND TRUCKS.

The subject of dynamos in general will be found in a separate volume of this series on that subject. The generators and motors described below are limited to such as are designated and constructed especially for railroad service, and therefore more properly belong under this heading of railways.

Generators.—The chief feature of generators for railways in which they differ from those for running lights is that they must be able to stand very much rougher usage and considerable overloading for short periods. They must therefore be designed on different lines than those for incandescent lighting. They must also have their wires more carefully insulated from the frame of the machines, because in railway service one pole is almost invariably grounded, which adds a greater strain on the insulation, which besides this is subjected to 500 volts instead of the usual 110. A feature worth noticing is that most of the new railway generators are made multipolar and many of them use carbon brushes.

Westinghouse 500 Horse-Power (370 kilowatts) Six-pole Generator.—In the accompanying illustration it will be seen that' in general design it resembles somewhat the well-known alternating current dynamo of that company. There is the same cylindrical yoke parting along a horizontal plane through the shaft, and the same arrangement of inwardly

FIG. 32.—WESTINGHOUSE GENERATOR, 500 HORSE-POWER.

pointing pole pieces. It is claimed to be the largest American built dynamo. The 125 and 250 horse-power machines have four pole pieces, and the 500 horse-power six poles. In this latter machine the shaft has a bearing outside of the pulley, which relieves it in a large degree from the bending strain, and adds to the rigidity of the armature. The pole pieces are built up of thin sheet-iron plates bolted together and cast into the cylindrical yoke. The field is compound wound, the shunt and series coils being wound side by side upon metal bobbins, which are then slipped over the pole pieces, and held in place by bolts. The bearings are self-oiling as well as self-aligning, and a great amount of bearing surface is obtained by the great length and large diameter of the journals. The armature is of the Gramme ring type, the core being laminated in the usual manner. The conductors are laid under the surfaces of the armature in oval perforations which are afterward milled into deep slots. It is stated that the electrical efficiency of these generators is from 94 to 96 per cent. at full load. They are wound for 500 volts, but they can be worked up to 550 or even 600 volts with the same armature speed, by throwing resistance out of the shunt circuit with the hand regulator. The action of the series coils is such as to cause a considerable rise in voltage as the current increases, the rise in voltage being sufficient to make up for any ordinary loss or "drop" in the line. The number of sets of carbon brushes corresponds to the number of pole pieces, and the alternate ones, being of the same polarity, are connected together. Each individual brush has its own spring, by means of which its pressure on the commutator may be adjusted, and any one may be removed without disturbing the others. The dynamos may be run in either direction. A 1,000 horse-power machine of this type is said to be in course of construction.

Thomson-Houston (250 kilowatt) Four-Pole Generator.—In general outline this machine, as will be seen from the adjoining figure, is quite similar to the large Oerlikon (Swiss) machines designed by Brown, and probably well known to many readers. It has four radial magnets connected around the outside by a massive octagonal yoke piece. The armature is of the Gramme ring type. In order that the conductors inside the armature may be held securely in place, an adjustable, internal wire support has been designed. When the armature is being wound the wires are forced into position so that they cannot sag, vibrate, or chafe the insulation. The commutator has 180 sections. The fields will be separately excited, although the connection at the switchboard is so arranged that by throwing a switch the dynamo can be made self-exciting should emergency require it. The movement of the brushes is effected by means of a shaft moved by means of a hand wheel, on which a small worm is attached, and which in turn works in a rack fastened to the yoke. By means of this a very fine adjustment of the brushes can be made. The worm locks the yoke so that it cannot be moved except by hand. The total floor space occupied by this generator is 13 feet 3½ inches by 7 feet 1 inch. The height of the machine is a little less than 8 feet. The pulley is 43 inches in diameter and has a 35-inch face. The speed is 400 revolutions per minute. The complete weight is about 21 tons.

One of the most important features claimed for this generator is the arrangement for lubrication and good alignment of the bearings. The boxes are made in two parts, and are entirely separate from the stands. On the top of the stand is a seat into which the spherical surface of the box fits, and in which the box is free to move. The bolts which secure it to the stand are smaller than the holes which are drilled through the box, so that a slight play of box in the seat is permitted.

The bearing shells or linings are removable, and are made in the following manner: A skeleton shell of brass is made, the interstices of which are filled with Magnolia metal. This is then bored and reamed to size, oilways being cut so that the oil circulation begins at the point where the oil rings touch the shaft. This method of manufacture permits a perfect circulation of oil, insures the cool running of the bearings, and greatly reduces the care and attention required by the dynamo when in operation. This type of box and bearing lining has proved so satisfactory that it is now being introduced in machines of smaller size, and will in future be used on all machines of large capacity. In order that the outboard bearings may be perfectly aligned with the other bearings, the stand of the extension base has two adjustments—one in a horizontal and the other in a vertical direction. These adjustments are made by the screws. Whenever it is necessary to examine bearing linings the armature is jacked up

FIG. 33.—THOMSON-HOUSTON FOUR-POLE GENERATOR.

about one-sixteenth of an inch, so that the bearing is relieved of its weight, two bolts removed from each stand and the entire box taken out. In case it is not desired to remove the box the cap can be taken off and the bearing linings readily removed.

Short (112 kilowatt) Four-Pole Generator.—The field magnet frame weighs over 800 pounds, nothing but the softest and purest iron being used in the melting pots. It is annealed very slowly in the molds, and when finished is so soft that it can easily be indented with a hammer. To this frame are bolted eight field magnets carrying the shunt and series coils and provided with the eight pole pieces making it a four pole machine. The general type of the frame is similar to the Brush, Schuckert, and Mordey machines. Upon a shaft nine feet long by six inches in diameter is keyed a spider carrying the foundation ring upon which the armature is built up. The armature is of the "flat ring" type and the core is formed of sheet iron wound spirally on the foundation ring and riveted firmly together. The outside circumference of the ring is somewhat wider than the remainder, and this portion is milled out into notches forming a modified Pacinotti ring. The coils are then wound on the core, the method being such that each one of the 200 coils is exposed to the air on all sides, thus securing ventilation. The projecting coils are a sort of fan, and in standing before the machine the current of air set in motion by the armature can be detected 10 or 15 feet away. As a consequence, both armature and field run cool, and there is therefore less likelihood of burning out a coil even with heavy overloads. It is stated that a burned out coil can be wound by any good mechanic at a cost of two or three dollars and a half day's labor. One of the features of the armature is its large diameter, viz., 36 inches. The armature shaft runs in large self-centring and self-oiling bearings, the lubrication being acomplished by rings carried by the shaft and drawing oil from a reservoir in the usual way. At the commutator box is also found an adjustable ball bearing thrust collar containing several hundred balls, and so arranged as to carry the armature thrust in either direction without heating. The commutator is quite large, being 20 inches of diameter; it has 200 bars, so that the pressure between two adjacent bars is only 5 volts. It is a four-pole machine, and there are four multiple carbon brushes carried by two independent collars and sets of brush holders. The compounding has been carefully calculated, and the "pressure curve" is a straight line, passing from 500 volts at no load to 525 volts at full load, with the speed maintained constant at 500 revolutions.

A very good illustration of this generator, too large to be reproduced here, will be found in *The Electrical World*, Sept. 5, 1891, p. 165.

Baxter Multipolar Generator.—This generator is designed for slow speed. It is an eight-pole machine, the field being built up of eight magnets, forming consequent poles. The magnet cores are laminated, and are clamped together so as to form a rigid ring, almost as much so as if they were made of one solid piece. This field ring is secured to the frame of the machine by four stout bolts attached to the poles of similar sign. The armature is a tooth "Gramme" ring, and the total air space is very small, being but a small fraction of an inch. The magnetizing force required to saturate the field in the 10 horse-power motors is less than 4,500 ampère turns. The shaft runs in two large bearings in the centre; the driving pulley is placed at one end, while the armature is at the other. The bearings are both in one solid

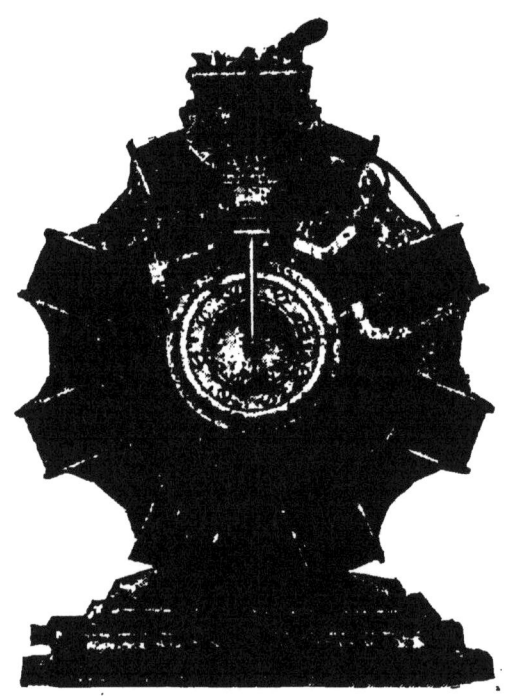

FIG. 34.—BAXTER MULTIPOLAR GENERATOR.

frame, so that it is impossible for them to get out of line. By this construction the machine can be easily taken apart. All that is necessary is to take off the pulley and the brush holders, then the armature and shaft can be removed at once. This design, with the addition of an outside bearing for the pulley end on the shaft, will be used in the large generators for railway work. They will either be compound wound or separately excited. The makers advocate separate exciting where it is desired to obtain the very best results. Although the speed at which the machines run is very low, their efficiency is said to be very high, and the output per pound of weight is claimed to be greater than can be obtained with the ordinary two-pole machines.

It is stated that the 8 horse-power 500 volt machines, which revolve at 800 revolutions per minute, will weigh complete a little over 500 pounds, and the electrical efficiency will be over 95 per cent. This is equal to about 12 watts per pound, which is very good. The 75 horse-power generator, which runs at 500 revolutions per minute, will have an electrical efficiency of over 96 per cent., and will weigh about 6,000 pounds, which is about 9.3 watts per pound. This generator has an armature 24 inches in diameter, with a 9-inch face. The shaft will be 4½ inches in diameter, and of hardened steel.

The accompanying illustration shows another form of the Baxter railway generator. It is wound for a potential of 500 volts, which is kept constant, regardless of load variations, by means of a separate exciter, and the multipolar field construction, while the Gramme ring armature, of large diameter, allows of a slow speed, thus reducing to a minimum the wear on the commutator, brushes, bearings, belting, and shafting. The brush holders are so designed that the brushes are automatically fed down upon the commutator with a light and constant pressure which insures long life to the commutator, and eliminates sparking due to uncertain pressure. A special feature of this generator is

FIG. 35.—BAXTER GENERATOR.

the mounting of the pulley between the bearings, thereby relieving the machine from all torsional strain.

Motors, Gearing, and Trucks.

The subject of motors in general will be found in a separate volume in this series on the subject of dynamos and motors.   The descriptions given below are limited to those of motors specially designed and built for railway service, and therefore more properly belong under the heading of railways.   The subjects of gearing and truck are so intimately connected with that of motors, that no attempt has been made here to separate them, except in a general way.

Railway motors have passed through such a stage of development during the past year that they may almost be said to have been completely revolutionized.   Double reduction motors, in which two pairs of gears were used, have given way to single reduction motors, having only one pair of gear wheels. Some makers have gone even farther and have introduced a gearless motor, in which the armature is directly coupled with the shaft of the driving wheels.   Although the latter have been introduced in a few cases, they must still be looked upon to a certain extent as experimental. On account of their slow speed, the size and weight necessarily increase, and an additional feature is introduced owing to the fact that irregularities or dirt on the track will cause the resulting hammer blow to be communicated to the armature, unless this is protected by some means.   Other features which have been overcome, or which are desired, will be seen from the descriptions of several motors given below.

In a report of a committee of the National Electric Light Association on a perfect street railway motor Mr. Everett gives the following opinions on what he thinks are the most important requirements for a railway motor:

"The motors, when first constructed, were altogether too light, both mechanically and electrically, but these difficulties are being overcome very rapidly, as well as the serious difficulty of too rapid motion, which swelled the operating expenses very largely in maintaining the parts and replacing the gearing. The best motors manufactured hereafter will be the most simple in the matter of the construction of the parts, and at the same time not consuming too great a quantity of electricity, so that in addition to being simple they will also be economical. I think it would be an improvement if in all machines made a better insulated wire were used in both armatures and fields. A more reliable and positive fuse wire application, one that would always burn out while still under the capacity of the motor, would also be a great improvement.

"I would again reiterate the fact that motors have been wonderfully perfected within the past year, and if as much progress is made during the coming year there can be very little to ask in perfecting a motor, although it is very desirable that the mechanical application should be more carefully looked after.

"All the prominent companies seem to have fallen into the same error, and seem to persistently and maliciously continue in their evil ways. I do not think that any one subject connected with electrical propulsion has received so much attention from the railways as this one, and with so little co-operation and assistance from the electrical companies. I, of course, refer to the price of their equipment. It appears to me that the companies, instead of operating 4,000 motor cars, could be operating 40,000 within a very short time, if they would bring the price down to a reasonable figure, so that all companies could afford to purchase an equipment. I think it would be desirable if the electric companies would supply all extra parts from their shops at a price allowing a reasonable margin for profit.

"The perfect motor ought to have, as hereinbefore suggested, a reliable fuse plug that will invariably blow before injury is done to the machine. It is desirable to use a controlling switch that is easily operated and readily reversed, in case of accidents. The simpler the controlling device the better, and it should be constructed with a view to guard against any possible disarrangements of the parts, so that it will be reliable in all cases, both electrically and mechanically. The rheostat should also be carefully looked after, and properly protected to keep it from injury, by reason of water, snow, or dirt getting upon it. It should only be available in starting the car to avoid the lunge of a start, and should be so arranged as to be cut out as soon as the car is started, and give the entire efficiency of the motor proper. The motor should be well protected in all parts from any outside interference, so that in running along the street it will be impossible to pick up nails, wire, or anything that would short circuit it, at the same time observing that a motor must be properly ventilated to keep it from heating while in use. The cover should be made so as to be easily removed.

"I deem it very advisable to have an armature of a large diameter, making a small number of revolutions per minute, with the bearings made of extreme width, with proper grease cups, and in such a condition that they can be readily re-babbited when slightly worn. The diameter of the commutator should also be large, and to have the brushes easy of access is very desirable. The winding of the armature ought to be of the simplest kind, and the size of the wire and insulation of same should be carefully looked after. I think the insulation wires in armatures is at present one of the weakest points in the motor.

"The armature gears should have a wide face, and run in oil. The armature shaft ought to be of ample diameter, and there is nothing gained by having the keyway too small for the securing of the commutator to the shaft. The commutator should be carefully insulated, so that there will be no grounds between it and the case. The box in which this gear runs ought to be constructed of copper, or some light material that is somewhat flexible, so that if struck from the outside it will bend rather than break. The fields should also be wound with a wire of better insulation, and of ample size to take the current. Of course, in this particular I do not intend that the wire of either field or armature should be great enough to take more horse-power than ought to be used by the machine. To my mind it is very desirable to have the armature in such a condition that it can be readily taken from the machine and put in again.

"One of the serious disadvantages to operators of

FIG. 36.—SHORT GEARLESS MOTOR.

electric roads is the expensive labor necessary in winding the armature and fields, also in regard to high-priced mechanics who ought to be employed to attend to the machines. There is nothing gained in employing a cheap class of labor to handle an electric equipment, either as electricians, armature or field men, or mechanics. This proposition is a self-evident truth, as can readily be observed in many roads now in operation. At present (October, 1891,) I think the single reduction motor is the nearest perfection of any on the market. The durability of the motor is a question which requires very careful attention. The single reduction motor, when properly looked after, ought to last for many years. We have had one in operation for over 10 months, and it appears to be in as good condition as when it first went on the road. The noise of the motors has been very largely done away with, and by careful attention the old countershaft machines can be used until worn out by simply covering the gearing with an oil box, and by not attempting to run them too many miles without inspection."

Gearless Motors.

Short Gearless Motors.—The accompanying illustrations, Figs. 36, 37, 38, show the method of mounting this motor. The armature, which is intended to run at from 100 to 150 revolutions per minute, is mounted upon a hollow shaft through which passes the car axle. This hollow shaft is made of steel, and is about six inches in diameter on the outside, with an opening inside of nearly five inches, so that there is a clearance between the axle of the car and the inside of this hollow shaft of an inch space all around. In the centre of this hollow shaft the armature and the commutator are fastened, and on each end of the armature shaft are keyed two heavy crank discs, made with an iron hub and rim, and a wooden web. This thoroughly insulates the armature shaft from the rim of the crank wheel. This crank wheel rim has upon one side a crank pin, as will be seen in the illustration. The car wheel has a crank pin also, and between the two is stretched a heavy coil spring, capable of pulling with a very slight elongation 2,500 or 3,000 pounds. The power of the motor in turning the wheel is transmitted through these springs. Just inside the crank discs are the bearings of the hollow armature shaft in the motor frame. This motor frame is made of two castings of steel with arms projecting forward and backward (see Figs. 37, 38, 39.) These arms rest on rubber cushions placed on channel bars which go from side to side of the motor frame. In this way the armature is held up in the frame and away from the axle, and the entire weight of the motor is supported by these spring cushions and by the car boxes outside of the wheels at the points where the car body is supported. None of the weight of the motor falls on the axle inside of the wheels. The spring cushions upon which the motor frame rests are limited in their movements to a fixed distance, so that the inside of the armature shaft cannot come down upon the axle. By these means the motor is carried entirely on springs.

A sheet-iron casing one-fourth inch thick covers the entire motor. This casing is so hung on hinges

that it can be swung open to allow examination of
the motor from beneath.   The motor is constructed
entirely of iron and steel, and its weight is about the
same as that of the standard type of Short motor,
and its efficiency is stated to be somewhat greater.

The armature has the form of a flat Gramme ring
wound with a very large number of independent
sections.   The diameter is as great as is consistent
with getting the motor under the car, giving, of
course, a powerful torque to compensate for the lack

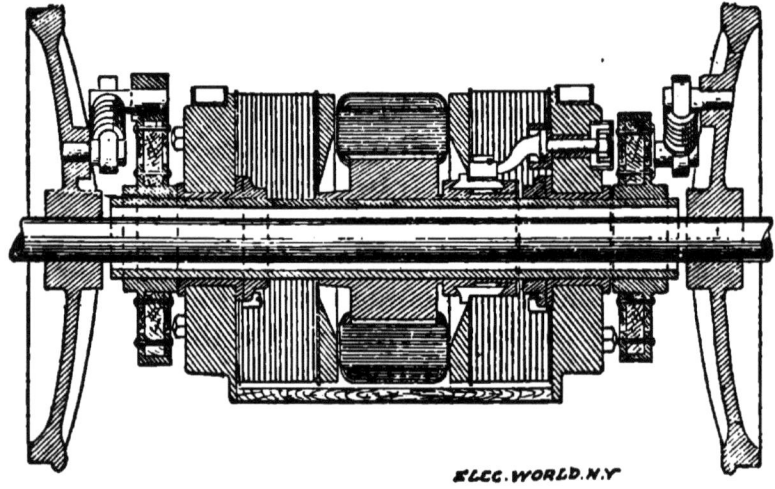

FIG. 37.

of gearing, and also a high back electromotive force
to secure as much efficiency as possible at low rotative speed.   The field magnets are of the regular
Brush type, eight in number, producing a four-pole
machine with a very intense field and very narrow
air gaps.   The first movement of the armature is
cushioned, as shown in the cuts, by a pair of
springs, one acting in tension and the other in com-

pression, and the full torque of the armature comes upon the axle only after a rotation of nearly 60 degrees. This feature of a semi-flexible connection between armature and axle is said to lessen the wear and tear on the armature very considerably.

Owing to the large diameter of the armature, a 36-inch car wheel is recommended for use with this motor, and under these circumstances there is a clearance of five and a half inches between the low-

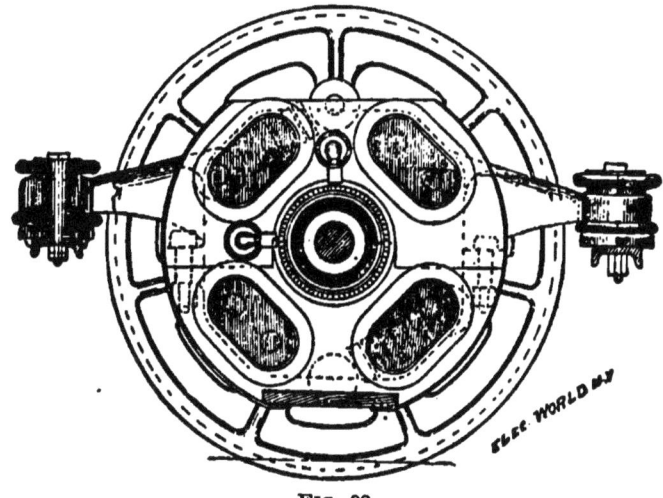

FIG. 38.

est part of the motor and the track, ample to allow for contingencies that are likely to arise. Whatever efficiency may be lost in winding for slow speed is claimed to be more than made up for by the gain of dispensing with the waste of power always incident to gearings. The resistance of the motors is an ohm and a quarter for the two coupled in parallel. In trials made in Cleveland the motor was subjected to the severe test of reversing at full speed without injury to the armature or producing any result more serious than a violent spinning of the wheels in the wrong direction. The 15 horse-power machine very closely resembles the 20 horse-power illustrated, except that it has only two poles and four field magnets, arranged as in the ordinary Brush dynamos; otherwise the construction is the same as that of the larger machine.

It is very difficult for any motor, gearless or geared, to crawl through city streets efficiently, and the gearless type is at a special disadvantage in this respect. Its best field is on comparatively level suburban roads, running at rather high speeds, and for such service it is capable of giving excellent results. The absence of heating and the type of armature employed—with widely separated coils very thoroughly ventilated—give good reason for expecting small repair bills, and even in case of accident damage is easily repaired, as a coil can be wound in a short time without removing the armature sleeve from the axle or doing anything more than clearing away the fields to obtain ready access to the commutator.

A later description contained the following note and illustration (Fig. 39): As yet, no roads have been fully equipped with this motor, so that a careful study of its performance under various conditions is still lacking. The result of a year of investigation and experiment on gearless motors by the Short company went to show that a working gearless motor would be unlikely to prove successful unless it was of multipolar construction, with very powerful magnetic circuit and small magnetic gap,

equipped with a Gramme armature of comparatively large diameter. These characteristics were
embodied in the present gearless motor, which in
its latest form is shown in Fig. 39. There are but

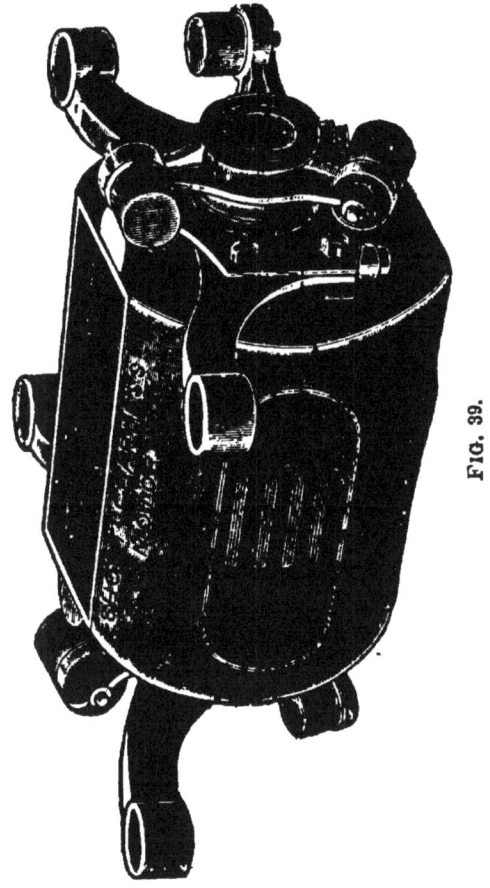

FIG. 39.

three wearing points on each motor, and the armature is of extra large diameter; there are eight field
magnets making four poles. A three-armed spider
is placed on each bed of the hollow shaft, each arm

is provided at the extremity with a socket to receive the rubber cushion or spring, these cushions bear upon lugs cast on the car wheels, and as the armature shaft and spider revolve the action is transferred to the car axle. The rubber cushion serves the double purpose of insulation and easy starting and has replaced the metallic spiral spring of the earlier forms of the motor. The insulation from the truck is very complete. The distance from the centre of the axle to the bottom of the casting is 12 inches, and at a speed of 12 miles per hour with 36-inch wheel the armature revolves at 94 turns per minute. The electrical output of the motor has been quite carefully investigated, and as an average of a large number of readings taken on the three electric lines in Cleveland the following results have been obtained: Average volts, 480; average ampères, 24; electrical horse-power, 15.44; average number of passengers, 48.

Regarding this gearless motor, Mr. C. C. Curtis, of the Short Electric Railway Company, said: "In the city of Rochester there has been kept day by day an accurate record of the motor and generator repairs. That road started running in November, 1890, and giving it eight months of run, through the winter months—the hardest months in the year—up to the first day of August, the average cost of repairs per car mile was four mills. When I say four mills per car mile, I mean only the repairs on the generators and the repairs on the motors; that is, the electrical repairs. In Muskegon, Mich., where we have been running about a year and a half, our record shows two mills per car mile. By removing four bolts, you take off the two lower wheels and raise one end of the car and you roll out your armature and the car axle. Should a bobbin burn out, it can, because of its peculiar construction, be repaired for from two to three dollars. These bobbins are only about three-quarters of an inch deep and surround the outer circle of the armature. They are not connected; and, should one burn out, it does not interfere with or necessitate the burning out of any of the others. The hue-and-cry has been raised that the gearless motor is a very nice thing, theoretically, but in practical operation it is going to take an enormous amount of current. I have a report of a test made in Cleveland, O., by Mr. Al. Johnson. It was a trial between one of his cars and one of the Short gearless motor cars. These two cars ran over the Brooklyn street railroad line, running about 20 minutes apart, doing commercial work. The gearless car checked up some 80 passengers, and the single reduction motor, which Johnson was running, checked up 47. The car ran for about two hours and a half, and we have the half-minute readings. The report and the readings show that the single reduction motor takes 24 per cent. more current than the gearless motor."

. Westinghouse Ironclad Gearless Motor.—As will be seen in the figure, this motor is completely surrounded and protected by the field frame, which forms a casing of sufficient strength to withstand all shocks and obstructions of the roadbed. The field consists of two symmetrical castings of special iron, sleeved upon the armature shaft or axle, hinged on top and secured together by bolts. The joints are made watertight, and the bearings are provided with leather cups for the same purpose, which makes it dust proof. The armature, which is of the drum type, is built upon the car axle. The sheet-iron discs, being solid and keyed to the axle. give

FIG. 40.—WESTINGHOUSE GEARLESS MOTOR.

the axle an additional strength, which precludes any possibility of its bending. This arrangement, of course, eliminates all gearing. The car wheels are fastened to the shaft by a new arrangement, which makes it possible to replace them easily and quickly without any special tools or skilled labor. The armature is but 16 inches in diameter, with a grooved periphery for the wires, which not only increases the efficiency, but holds the wires rigid. It is securely fastened to the shaft, and connections with the armature are made by short, heavy wires. The brush holder, which is rigidly fastened to the magnet frame, is well insulated and easily accessible by openings provided with watertight lids. The weight of the magnet frame is counterbalanced and cushioned upon powerful spiral springs which rest upon the cross bars of the truck. These springs prevent the field from rotating, and give the motor the necessary flexibility for easy starting. The total depth of the motor is but 20 inches, giving 5 inches clearance between the bottom of the motor and the rail, with a 30-inch wheel. Actual tests are said to have shown a working efficiency of 90 per cent. It is also claimed that, after two hours' run with a load of over 20 horse-power the rise in the temperature of the armature and field coils was only 30° C. above the surrounding air. There are three types of gearless motors made; one for heavy grade city work, one for ordinary level city lines, and a third for suburban service.

Dahl Slow Speed Motor.—This motor is not strictly speaking "gearless," as it drives by means of a friction clutch, but as there are no reduction gears, it properly belongs to this class of gearless motors.

In this motor an armature of the ring type of large diameter is attached to a non-magnetic spider, which is placed on a shaft or on the car axle, and is

free to revolve about it. Attached to this at one end
is one-half of a friction clutch, the other side of
which is attached to the axle by means of a coiled
spring. On each side of the web of the spider are
two magnetic spools and cores, through the latter
of which the hub of the spider is free to revolve.
Consequent poles are formed by these magnets on
the interior and exterior of the armature, alternating in position, those magnets on the outside being
brought together and held in position with dowel
pins, and a yoke being bolted rigidly to them. This
yoke is fastened to the frame or truck to keep the
field magnets from revolving. The whole motor is
protected from dust and other foreign substance by
a brass case arranged to be accessible. All fuses,
connections and brushes are at the top of the motor,
and are accessible to the operator when he has
opened a trap-door in the car and one of the slides
in the casing. The clutch, which is one of the features of the motor, is composed of sheet-iron discs,
each alternate one being fastened at its centre or
periphery. Those fastened at the centre are attached
to the end of the spider of the armature. Those fastened at the periphery are joined to a spring, as
before mentioned. Rigid action is secured by pressing the sheets together by springs interposed in such
a manner that at a predetermined pull on the clutch it
slips and allows the armature to revolve faster than
the axle. The armature is so constructed that each
individual section can be replaced without disturbing
the others, should it be needed. By the use of the
clutch mentioned it is claimed that it will be impossible to overload the motor beyond a predetermined

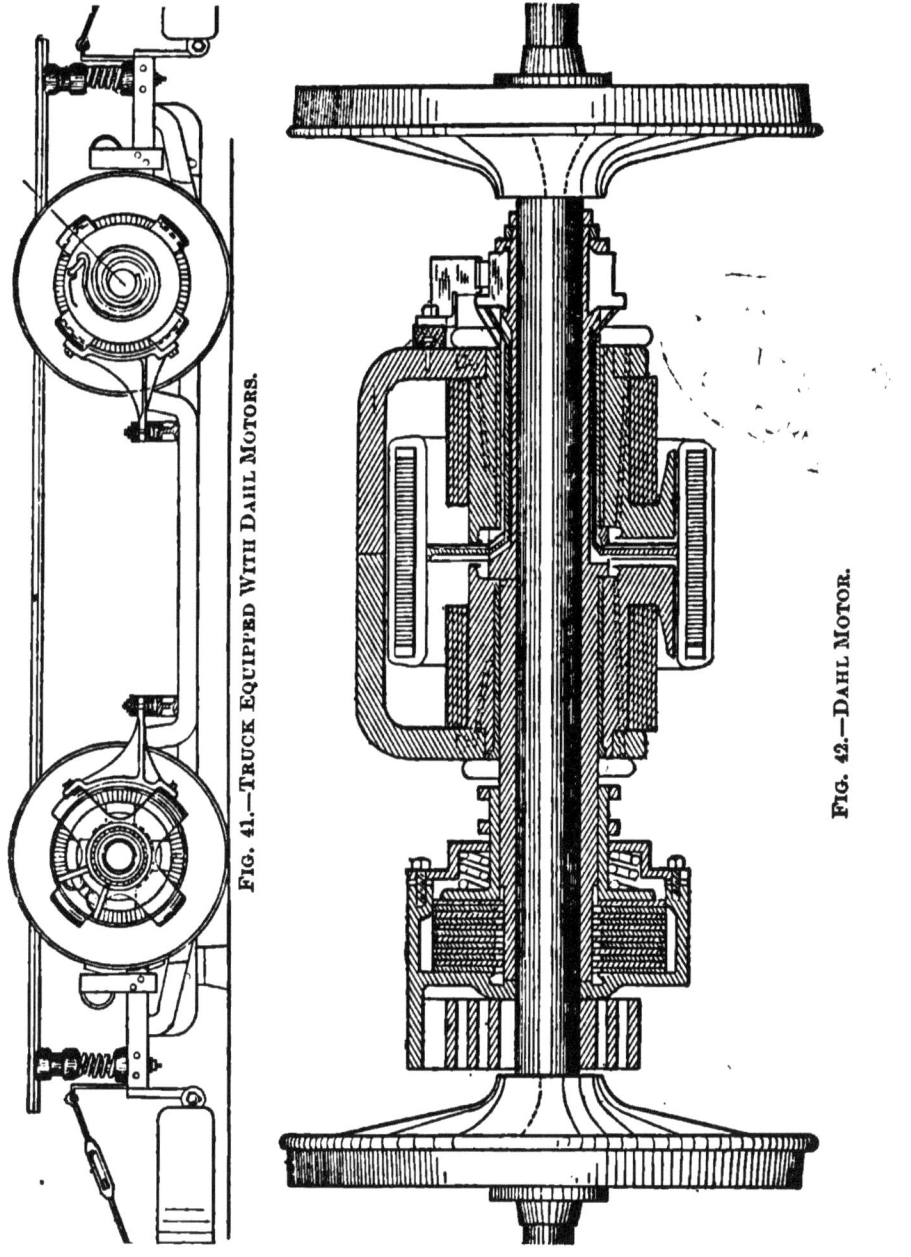

FIG. 41.—TRUCK EQUIPPED WITH DAHL MOTORS.

FIG. 42.—DAHL MOTOR.

point. The armature may be kept running continuously, thereby gaining whatever advantages such a method of operation may possess. The construction of the motor is such that it is practically ironclad, with no outside leakage of magnetism. Mitis iron is used for magnet cores, and the clutch is composed of 40 discs of iron. The weight of a motor rated at 15 horse-power is 1,690 pounds.

Eickemeyer Motor and Truck.—This form of motor and truck embodies the features of both the gearless and reduction gear motors in a combined motor and truck. Fig. 43 shows the truck complete, as fitted with the motor; Fig. 44 gives an idea of the various parts and their construction, the motor being removed from the car, its frame inverted, and the armature, coils, etc., shown separately. A drum form of armature is used, having 74 coils. These are wound on an arbor, from which they are removed and thoroughly dried and insulated, and then placed in position so that, should a coil become damaged it may easily be removed and replaced without interfering with the other coils or with their work. The armature is supported midway between the two axles and has its shaft connected by ordinary connecting rods with crank pins to both axles, the connecting rods attached to both ends of armature shaft, and the cranks set at an angle of 90 degrees, preventing the armature from ever getting on the dead centre and also giving a maximum starting torque in all positions. A second feature is the use of only a 26-inch car wheel in place of the usual 36-inch wheel required to raise the gearless motor above the roadbed, the smaller

wheel allowing a higher armature speed, and consequent increase in commercial efficiency. Owing
to the form of frame no extra seating is required to
protect the motor, as it is entirely inclosed in an

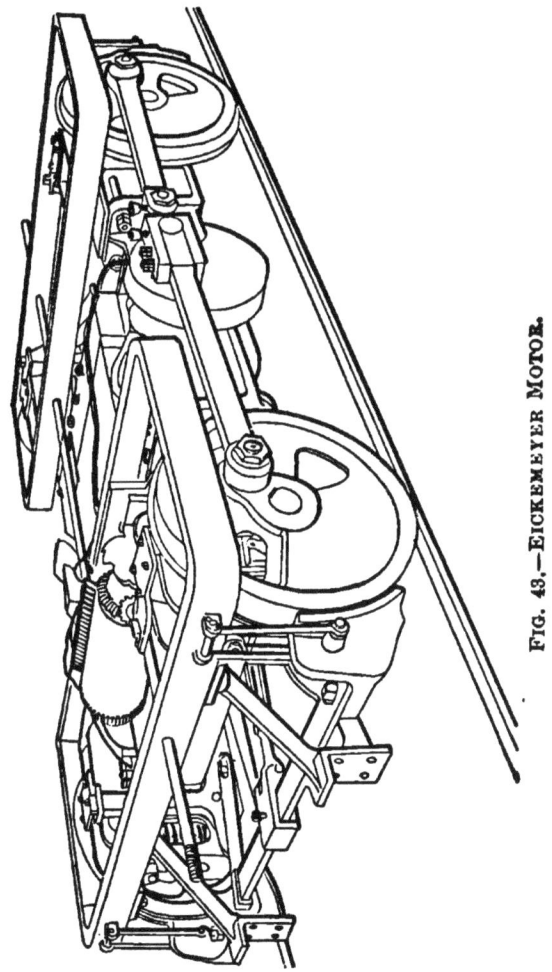

FIG. 43.—EICKEMEYER MOTOR.

ironclad casing. The rheostats are placed in boxes
supported over the axles, and the controlling mechanism connected by shafts running the entire length

of the car may be operated from either end; while all switch mechanism is protected in casing below the car floor within easy access for repair and inspection. The sills of the car body rest on the four iron brackets placed on the motor castings, and on beams that extend across each end of the truck.

Leonard's Method of Avoiding Gearing.—Under the subject of motors, in the volume on "Dynamos and Motors," will be found described a method for running motors at any speed with constant torque and nearly constant efficiency. It consists essentially of using three motors and dynamos in place of one, so arranged that the current and voltage for the last motor may be varied to suit the requirements of the speed and torque. The portion of that description referring to railway motors is reprinted here in full, in the language of the inventor:

"For operating an electric railway we will place a shunt-wound motor on the car, and directly driven by this motor will be a special generator, which will be connected to the electric motor below the car. It is evident that the generator and working armature may be wound for any voltage desired, say 20 volts, which will make the problem of insulating the street-car motor an extremely simple one. If desirable, we can supply several cars of a common train from one special generator on the forward car. With this outfit we will be able to take any car up any practicable grade or around any curve with no more power than is required to move the car on a level, and always consume the same power, regardless of weight, grades, or curves. That is, the automatic increase of current, to take

FIG. 44.—EICKEMEYER MOTOR.

care of any increased torque, will be compensated
for by a corresponding decrease in the volts and
speed. We may start a car up any grade or curve
with but a small fraction of the power required for
normal speed on a level.

"I wish to call attention to a very important
development leading out from this, namely, that we
will be able to use alternating currents for operating our street cars, for it is well known that the
ordinary alternating current generators will operate perfectly as motors, if the speed and torque be
kept constant. Since by this system we can, from
a constant torque and speed, get any other torque,
and automatically a corresponding speed, we shall
be able to run street cars perfectly by alternating
currents. This, again, will enable us to dispense
with trolleys, conduits, storage batteries, etc. We
will place between our tracks, in manholes, converters whose primary pressure can be anything
required for proper economy, and whose secondary
will be say, 15 volts. This secondary circuit will
connect directly with the rails. The road will be
divided in sections, each a few hundred feet long,
and each section will be supplied by its own con-
verter.

"On first consideration, the additional apparatus
necessary would seem to make the system prohib-
itory in practice; but the capacity of the present
single motor is greater than the combined capacity
of the apparatus this system would require, and the
capacity of the prime motor is very much reduced.

"In order to reduce the first cost to a minimum
and yet secure the advantage of different automatic

speeds and high efficiency, I have devised two modifications of the arrangement described above. The
first is adapted to power in which a smooth, efficient
acceleration of·a load from rest is required, as in
the case of passenger locomotives and elevators.
The second case is where various automatic speeds
are desired, but no special importance attaches to
the starting of the load from rest, as is the case in
machinery in general.

"For the first case, we have the trolley system of
electric street cars as the most important. Let us
suppose we have two motors of 15 horse-power each
for the car. We find that for full speed upon a level
we require about 15 ampères at 500 volts. Upon
heavy grades we find that about 50 ampères are
required, and, as before, we have 500 volts. With
this consumption of energy we find that we get a
speed upon the heavy grade which is about one-
quarter of the speed upon a level. In order to operate upon my system, let us place upon the car a
motor generator, the motor part of which is wound
for 500 volts and 12½ ampères and the generator
part of which is wound for 125˙ volts and 50
ampères. The fields of the motor and generator part
are distinct, and are wound for 500 volts, as are the
fields of the two propelling motors under the car.
All these fields are supplied from the 500-volt trolley
circuit. In the field of the auxiliary generator is
placed a rheostat.

"Now suppose the car at rest upon a grade. The
motor generator is running, but the generator has
a very weak field. Its armature is connected by a
controlling switch to the propelling motors. We

now gradually cut out resistance from the generator field circuit, and finally get about 20 volts at the
brushes of the generator. With this E. M. F. we get
sufficient current to produce 50 ampères through the
armatures of the propelling motors in a saturated
field. This gives us the full torque, and the car
starts at a speed of perhaps half a foot a second.
This speed can be maintained constantly and indefi-
nitely, and the consumption of energy will be less
than two horse-power. This is less than three ampères from the trolley line. In practice, however,
the speed will be rapidly but gradually accelerated,
until we have 125 volts upon the terminals of the
propelling motors. We will now be running at one-
quarter speed, and will be consuming 125 volts and
50 ampères, that is, 6½ kilowatts instead of 25 kilowatts to get the same result with existing motors.
To put it another way, we will not be using as much
energy as is represented by the 500 volts and 15
ampères necessary for full speed on a level.

"The next step on the controlling switch will dis-
connect the armatures of the propelling motors
from the auxiliary generator and put the two armatures in series across the trolley line direct. We will
now go at a speed represented by 250 volts, that is,
one half full speed. The next step of our switch will
place the two armatures in multiple across the 500
volts, and the next and last step will place the 120-
volt auxiliary generator in series with the main
central station generators and give us 625 volts on
our armatures and correspondingly increased speed.
We will be able to go up a grade of six to eight per
cent. at full speed, with 50 ampères and 500 volts,

which, with the present motors, gives only about one-quarter of that speed.

"Under this arrangement it will be noticed that the only apparatus which could be called additional is the small motor of 500 volts for the generator part of our motor generator, which is useful, not only for starting, but for full speed also. In stopping the car we have an electric brake action delivering back energy to the line at full efficiency and not through a rheostat, as at present.

"If we have a train of, say three cars, so that we have six motors, we can start from rest with sufficient smoothness by placing all six armatures in series, which will give us something less than one-sixth speed as the first step. Then we can place three in series with two multiples, which gives us one-third speed. Next, two in series with three multiples, which gives us one-half speed; and finally, all in multiple, which gives us full speed. Under such conditions, we can dispense with the small converting plant altogether."

Single Reduction Motors.

Westinghouse Four-Pole Motor.—Fig. 45 shows a general view of the new motor closed in its casing and ready for use. Its general form is cylindrical, giving both the shortest possible magnetic circuit and maximum strength with minimum amount of material. All the sharp corners that tend to leak magnetism are eliminated, and the machine is claimed to be rendered thereby slightly more efficient. The details of the magnetic circuit are best shown by examining Fig. 47, which shows the casting freed from armature and coils and opened up to exhibit its arrangement. The

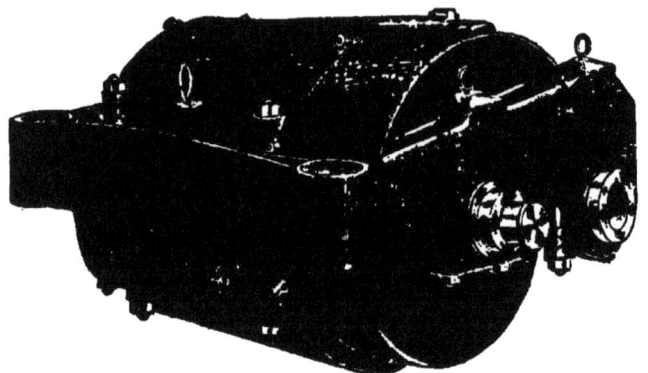

FIG. 45.—GENERAL VIEW OF MOTOR.

form of the magnetic circuit makes it possible to utilize four poles with great advantage; they are, as will be seen, rather narrow, and consequently are capable of being magnetized by comparatively

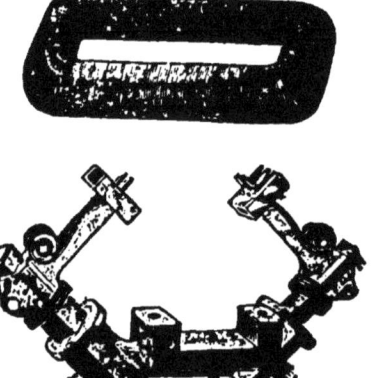

FIG. 46.—BRUSH HOLDER AND COIL.

short and small windings. One of these coils, together with a brush holder, is shown in Fig 46.

The brush holder is a casting bolted on to the lower
side of the main frame of the motor, and lifting its
brushes quite on to the top of the commutator,
where they rest 90 degrees apart. The castings are
of a specially soft grade of iron that has proved
to have excellent magnetic properties.

The gearing, inclosed as it is in an oil-tight case

FIG. 47.—MOTOR WITH ARMATURE REMOVED.

(Fig. 48). is always thoroughly lubricated and free
from dirt. All the bearings are bushed with metal,
and the armature shaft is tightly tapered to facilitate the removal of the pinion. The gear ratio is
3.3 to 1. The iron-clad form of the motor enables it
to be completely shut in by applying side plates,
so that in actual practice it is inclosed so tightly as

to be quite free from the numerous difficulties so often experienced from dirt and moisture finding their way into the working parts of a machine. As the lower surface of the motor presents only a solid casing it cannot be injured by casual blows from projecting rubble, a source of difficulty with which electric street railway men are only too familiar. Fig. 49 gives a perspective view of the motor, show-

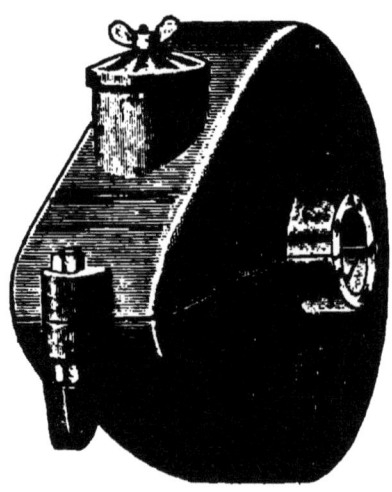

FIG. 48.

ing its arrangement in the frame and connection to the gears. The armature of the new machine is of the drum type. The core is built up of grooved iron plates, so that the windings are in slots upon its surface, thus completely imbedded in insulating material. The surface of the finished armature is therefore entirely smooth and the clearance space very small. Even should the bearings become worn so that the armature would brush against the pole pieces, no serious damage would be done because no wire is exposed.

The electrical efficiency of the motor is said to rise to 95 per cent., and the commercial efficiency to 75 or 76. Inasmuch as the efficiency of the two-pole motors of various makes with the complicated gear is generally held to be a little over 60 per cent., the abolition of the intermediate gear ought certainly to be good for more than 10 per cent. increase in efficiency. The normal speed of the armature at a car speed of about 10 miles per hour

FIG. 49.

is 380 revolutions per minute. The commutator is designed with special reference to obviating the heating that is sometimes so disastrous in street-car motors. Each segment of the commutator has a bearing along its entire lower edge, so that even if there should be any slipping the symmetry of the commutator would not be destroyed. The cross-section of the armature enables the two brushes, as before mentioned, to be placed 90 degrees apart, and both upon the top of the commutator, where

they can be readily inspected or replaced if necessary.

Thomson-Houston Slow Speed Motor.—The single
reduction gear motor, ordinarily called the "S. R.

FIG. 50.—THOMSON-HOUSTON SLOW SPEED MOTOR.

G.," seen in Fig. 50, is very nearly iron-clad, having
two pole pieces of ample surface and carrying two
field coils, which partially surround the armature
core. The magnetic circuit is completed on the front

end of the motor through the face plate and at the back through the frame on which are cast the axle boxes and arms which serve as a support for the armature shaft bearings.   The armature is of the Gramme ring type, and the bobbins are wound close together around the entire rim.   One advantage of this construction is the fact that any coil can be easily rewound without disturbing the others, while with the drum armature formerly used the windings all had to be renewed down to the injured coil. The brushes are placed exactly opposite and in a horizontal fixed position.   There seems to be no sparking under the ordinary running conditions, and the brushes are easy of access.   The field spools are protected on all sides by the fields and frame. The gears are entirely inclosed in a dust and oil-tight case, which is provided with a hand-hole closed by a spring cover, permitting ready examination of gears and the introduction of lubricants.

A sheet-iron pan extending above the centre of the armature shaft entirely incloses the bottom and sides of the motor and protects the armature and commutator from dust, snow, and water.   This pan has a sliding bottom, and is attached to the motor in such a manner as to permit of being readily removed for access to the various parts.   The motor when mounted on a truck with 30 inch wheels is designed to clear the tops of the rails four inches. The spur gear on the armature shaft is of steel, four and one-half inches face, and has 14 teeth.   The split gear on the car axle is of cast iron, with the same width of face, and has 67 teeth.   The speed of the armature shaft relative to that of the car axle is nearly as 4.8 to 1; when the car is running ten miles per hour the armature makes 538 revolutions per minute, or the speed of the armature is 53.8 turns per minute when the car speed is one mile per hour. The gears are surrounded by an iron box so that they may be run in oil.

The facility with which the armature can be removed simply by lifting the upper field, the ease with which an armature bobbin can be rewound, less liability to damage from centrifugal action, such as bursting of binding wires, displacement of coils, breaking of commutator connections, all insure a minimum amount of expenditure for repairs.

The accompanying illustrations show very plainly the general appearance and detailed construction of the motor. The advantages claimed by the designers have been summarized as follows: Noiseless single reduction gears, protected from dust and run in oil, and noiseless commutator. Electrical and mechanical simplicity attained by use of two-pole type of motor. Slow speed and powerful torque obtained by proper proportions. Protection for fields and armature from dust and water. Accessibility to all parts. A ring armature not likely to become damaged, and, if accidentally injured, permitting easy repairs. Moderate weight. Reasonable cost. Commercial efficiency. Small maintenance expense.

Thomson-Houston W. P. Motor.—The accompanying illustrations show a new motor known to the trades as the W. P. motor, which means water proof, because of the particularly complete iron-clad character of the field magnets.

It is a two-pole machine, based on the theory that the comparatively slight gain in weight efficiency that could be obtained with a multipolar type is more than offset by the increased complication of the windings. The only portions of the machine open to the outside air are exposed at the two oval openings at the ends of the armature shaft, and even these can be easily fitted with cover should it be desirable. The whole magnetic circuit is com-

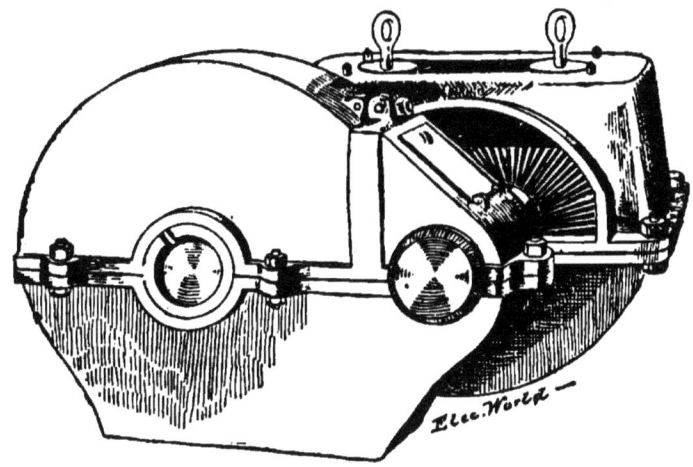

FIG. 51.—THOMSON-HOUSTON W. P. MOTOR.

posed of two castings bolted together and free to swing apart by a hinge allowing ready access to the armature. Fig. 52 shows the internal arrangements. The armature itself is very nearly twenty inches in diameter, a very powerful Pacinotti ring nearly six inches on the face and of about the same depth. It is wound with comparatively coarse wire in sixty-four sections, with fourteen turns to the section. Each coil is tightly placed in the space between two of the protecting teeth, and about the interior space the separate coils are closely packed, leaving only sufficient room for the four-armed driving spider. As will be seen, the armature takes up most of the full height of the machine, the pole pieces being but trifling projections and the requisite cross section of iron being obtained by extending the poles to form a closely fitting iron box that appears in the exterior view. An unusual feature

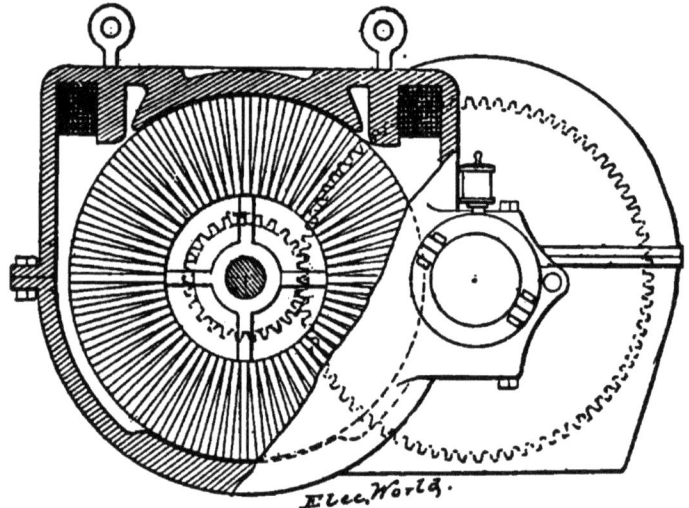

FIG. 52.—THOMSON-HOUSTON W. P. RAILWAY MOTOR.

is the use of but a single magnetizing coil wound not directly about the upper pole piece but on the casing immediately surrounding it. The lower pole is but slightly raised and both pole pieces have the greatest surface permissible with the dimensions of the machine. The use of a single magnetizing coil produces naturally an unbalanced field and a strong upward pull on the armature tending to relieve the pressure on the bearings. The iron-clad form, however, tends to distribute the lines of force so as to avoid the sparking and change of lead that might otherwise have to be feared. The single field coil is wound with quite coarse wire and its position is

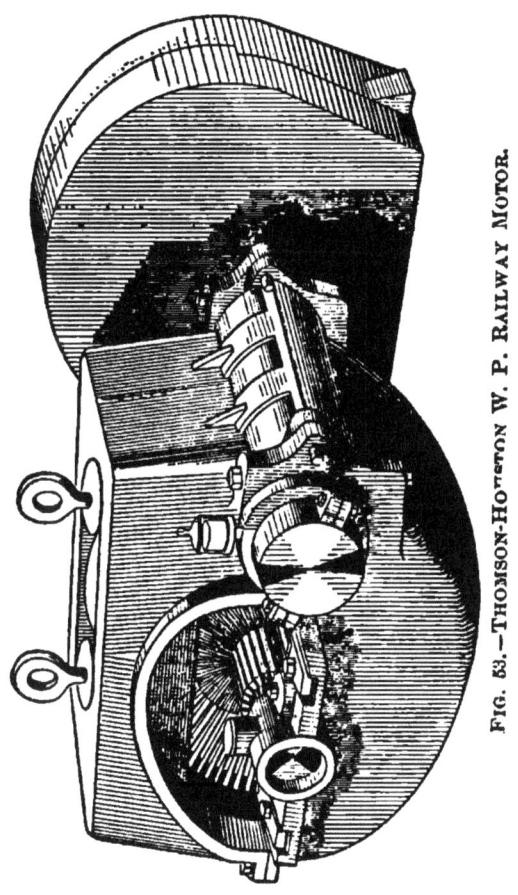

FIG. 63.—THOMSON-HOUSTON W. P. RAILWAY MOTOR.

claimed to insure the maximum magnetic effect from the current. The speed of the motor is about the same as that of the older S. R. G. form, but its general working efficiency is somewhat better, owing not so much to a greater maximum of efficiency as to a better working curve—at both heavy and light loads. The brush holders are shown in Fig. 53, and the slots in which they fit render their position evident. The brushes are of the ordi-

FIG. 54.—SHORT SINGLE REDUCTION MOTOR.

nary carbon description and are readily accessible through the opening at the end of the shaft. The gears run in oil.

Short Motor.—In this motor, known as the water-

tight, and more familiarly as the W. T., one pinion
and one gear have been dispensed with and the remaining gear is run in oil. The figure gives an idea
of the arrangement. It is claimed to be of about the
same efficiency as the other types, and is the lightest
and smallest of their standard motors. It weighs all
told a trifle less than 1,800 pounds, is incased and
completely protected by its iron frame and can be
operated on 30, 33 or 36 inch wheels, and any gauge
of track down to three feet. It is especlaliy recommended for narrow gauge roads, for mining and
similar purposes. Two sizes are made, 15 and 20
h. p. One of the features of these is the facility
with which repairing, in cases of necessity, can be
carried on. With the geared motors it is not necessary to remove the motor and car wheels; the car
can be run over a pit and every part of the motor
reached without difficulty. The armature coils may
be rewound without removing the armature from
the axle, and the field coils can be quite as easily
repaired. The commutator may be reached and
cared for with ease while the machine is running.
The motor field coils and armature are easy of
access, and the armature can be removed in case of
necessity by two men in eight minutes, its weight
being only 198 pounds. In fact, every part of the
motor can be removed without taking the machine
to pieces. On the Rochester (N. Y.) railway during
eight months the total cost of repairs, including
material and repairs to electrical machinery, was
but four mills per car mile, and the average loss
per car from necessary repairs was but four per
cent. of the total car mileage.

The Edison motor shown in the adjoining figure was illustrated but not described. It will be seen that it is a four pole frame, with only two coils; it is partially iron-clad. The armature is a Gramme

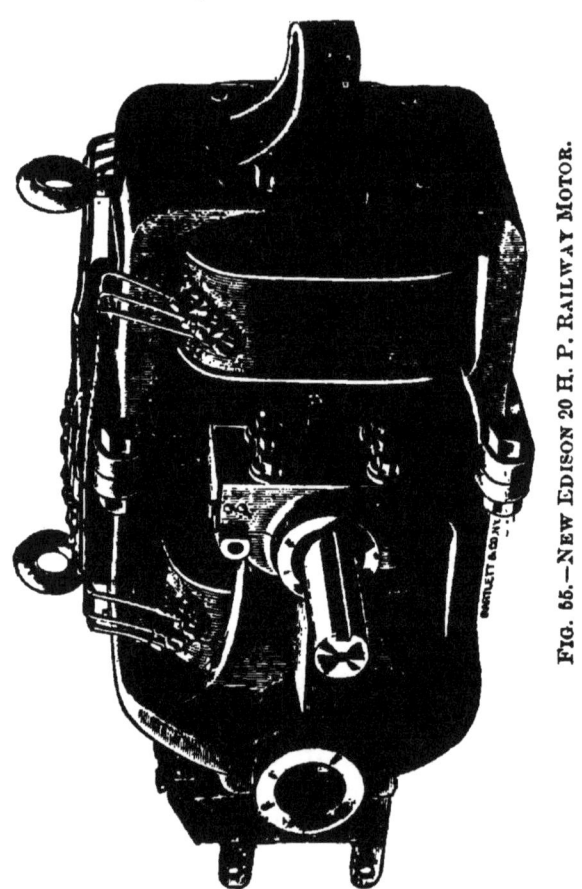

FIG. 55.—NEW EDISON 20 H. P. RAILWAY MOTOR.

ring, presumably with iron projections. It has a single reduction gearing.

A Swiss Motor and Truck.—The truck and motor shown in Figs. 56 and 57, used on the small Swiss road

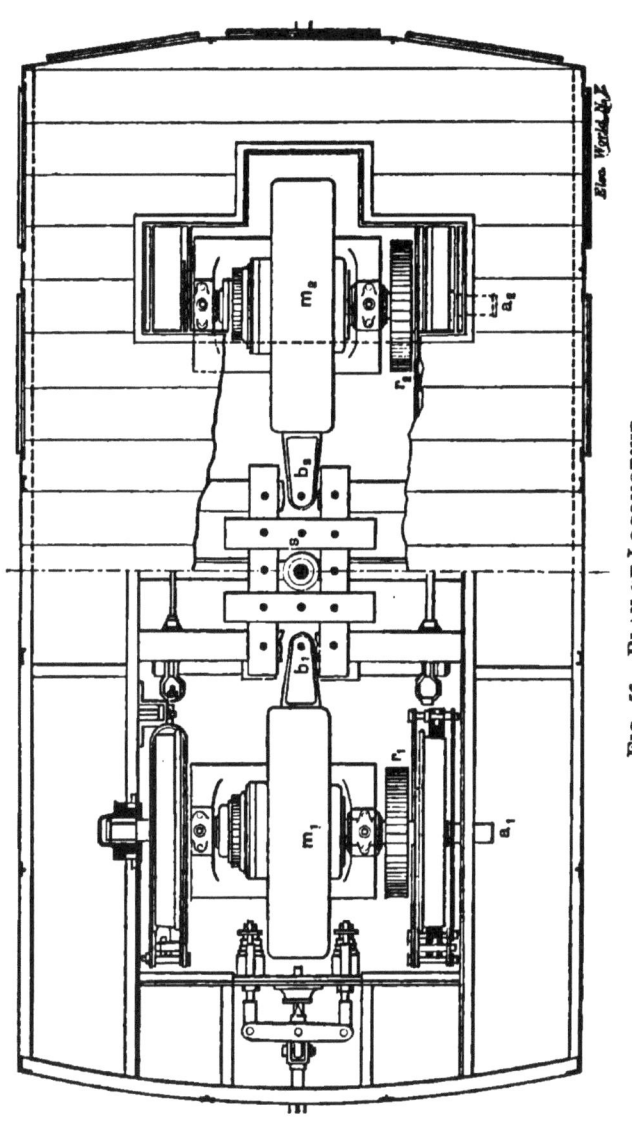

FIG. 56.—PLAN OF LOCOMOTIVE.

from Sissach to Gelterkinden, described elsewhere, are of interest as showing the practice in that country. The published description was very meagre, but the cuts will explain themselves. The motor was made by the Oerlikon Company and is their new standard type, they having abandoned the worm wheel motor, used before.

The construction of the electric locomotive is shown in Fig. 56.

The most important feature of the construction is the arrangement of electric motors, which rest directly upon the car axles, a^1 a^2. Power is transmitted to the car axle through a single set of gears. The motor armatures make about 450 to 500 revolutions per minute when the train is running at its normal speed. The placing of the armature shaft and car axle in the same vertical plane is claimed to give better results than the arrangement in which the four shafts of the armatures and car wheels are placed in one horizontal plane, since under all conditions of load the same distance is maintained between the centres of the intermediate gear wheels.

The motors, Fig. 57, are four-pole drum armature machines having a normal output at full capacity of 25 h. p. Each motor has but a single pair of carbon brushes which press against the bronze commutator in such a position that the motor man can reach them easily even when the car is in motion. The machine has been so carefully designed that the position of the brushes remains constant under all variations of load.

Siemens & Halske Motor and Truck.—The form

of motor and truck used at present by this firm is shown in the adjoining figures. This firm is the

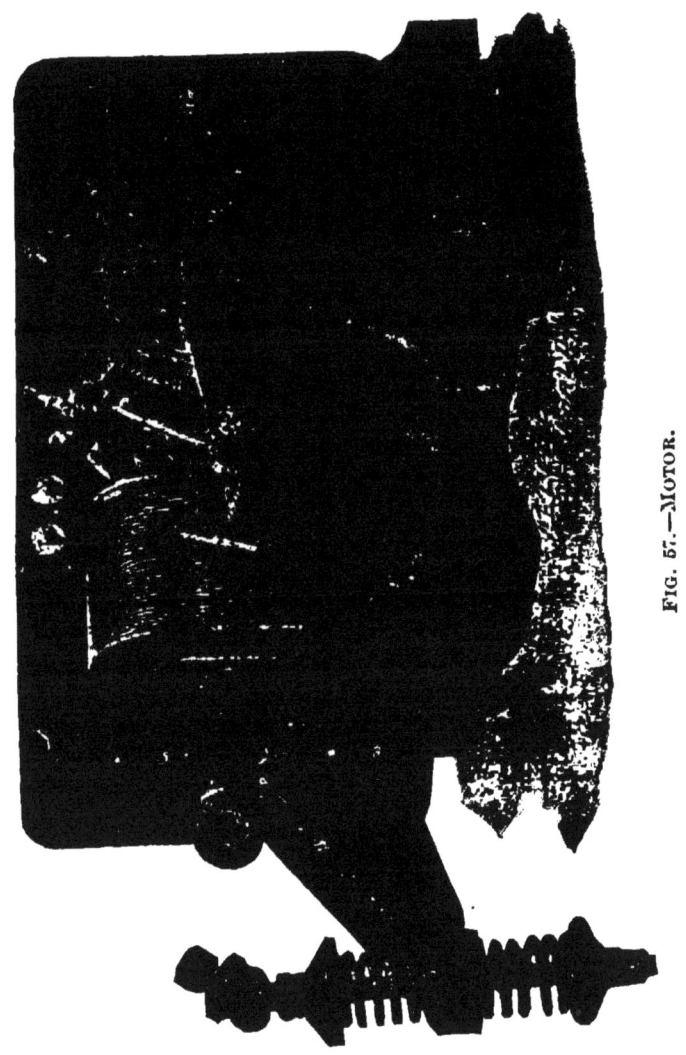

FIG. 57.—MOTOR.

largest and oldest electrical manufacturer in Germany and it will be remembered that they were the pioneer electrical railroad builders. They have furthermore constructed the only large plant of a conduit road running at present in the city of Budapest in Hungary. The motor and truck described below are those used on this line. In their trucks there is only one motor of 15 h. p. driving one pair of the four wheels with fixed axles; the wheels have a slight axial motion. There is a double reduction gearing from the armature to the axle; the first pair is driven by a chain to reduce the noise, the second pair sometimes consists of two gear wheels and sometimes of a chain. Fixed carbon brushes are used. The magnets are of wrought iron and the whole motor and gearing is inclosed in a zinc box.. The second pair of wheels, which are not driven, are connected to the motor frame by means of a flexible coupling, which enables them to turn. Such trucks can readily go over curves of 12 metres (39 feet) radius. The trucks are complete in themselves, and can readily be secured to the bottom of any car which has previously been used for horse· traction. They use 300 volts and state that it takes about 80 ampères to start a car for 24 passengers.

Henry Motor and Gearing.—Nearly all the present forms of street car motors are centred upon the axle and are free to move through a small arc about its centre. In this form, however, for simplicity and strength of construction, this method is abandoned for a design which reminds one of the English locomotive designs; the mechanical difference between this new motor truck and the usual ones is like that between the English and American locomotives. The latter, as is well known, was

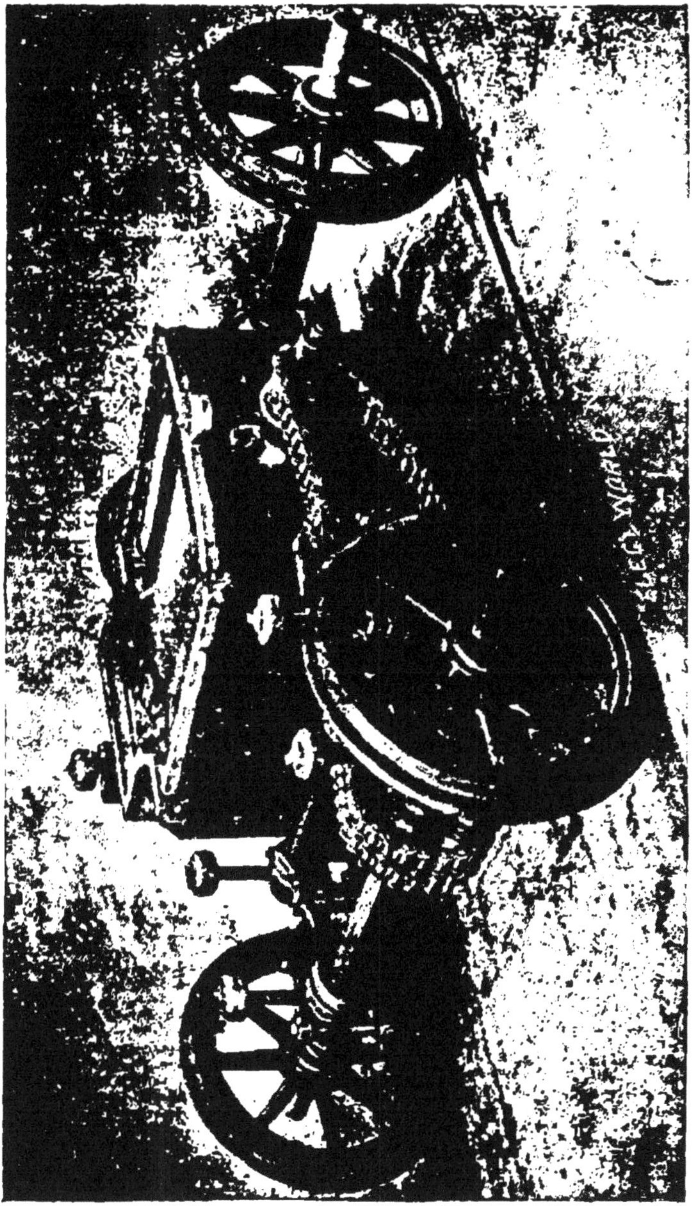

FIG. 58.—SIEMENS & HALSKE MOTOR.

designed for as great a degree of flexibility as the conditions of the problem allow, supported on springs at every point practicable. The English designers, on the other hand, rely on the perfection of the track for freedom from shocks, and make the frame work of the machine as strong and simple as possible. As might be expected the results of the two types are equally good under the two conditions for which they are severally intended. In the Henry motor and truck all spring supports arc deliberately avoided, and an attempt is made to reduce the working parts to the smallest number possible, and to support them so firmly that there will be no danger of their getting out of order.

Only one motor is employed, instead of two, and only one pair of wheels is driven by the gearing, the other being operated by a connecting rod. This construction, which might be objectionable on an old horse car track full of inequalities and continually getting out of line, is said to be excellent when used in connection with proper track construction. The motor itself is rigidly fastened to the steel frame that connects the axles, its ends having bearings on three-inch steel axes, parallel with the axles of the wheels. The magnetic circuit has but two joints, and its form is not dissimilar to that which has been employed by Reckenzaun in England. It is very short and the windings are very compact; it has two consequent poles that embrace a Gramme armature 18 inches in diameter.

The armature is wound with two layers of No. 7 wire divided into 72 armature segments, each connected to its appropriate commutator bar. The

FIG. 59.—THE BUDAPEST ELECTRIC TRAMWAY CAR.

winding is continuous throughout, loops being taken out at the commutator instead of the wire being cut. The face of the armature is 13 inches wide, and the clearance space is as small as is consistent with good mechancial construction. The insulation of the armature wires is exceptionally thick and firm, almost equal in thickness to that on line wire, and the whole is so firmly put in place in case of an armature section burning out it need only be cut off at the commutator, and the machine may then be run until the damage can be repaired. The fields are wound with No. 9 wire, and the four coils arranged for three combinations at the switch-board, all in series, two in series and two in parallel, and all in parallel, therefore giving three speeds by this rearrangement. It will be seen that the size of the wire used in this motor is much larger than it has been customary to employ, so far, and the resistance of the machine is therefore very low, so low, in fact, that it would be difficult to employ the machine without danger of excessive starting currents were it not for the clutch device through which it drives the truck. This consists in the well known epicyclic gearing arrangement, capable of running freely when the epicyclic pinion is free to move, and exerting its full power when the pinion is held fast by the clutch lever. The armature therefore can run all the time, or be allowed to give its full speed before any of the load is thrown on ; as soon as the clutch is tightened, it can take hold and utilize its momentum in starting the car, thereby avoiding a sudden rush of current that has proved so disastrous to many armatures. The epicyclic gear is inclosed in a tight case and runs in oil, so that the noise is very much diminished, and the wear is much less than with exposed gears. It will be noticed that the armature is geared down but once instead of twice. The reduction is seven to one instead of twelve to one as in many other trucks. A slight modification will allow the armature to be connected directly with the internal gear after the car is under way, thus allowing very high speeds. A mechanical addition to the motor gear is an automatic brake; there are two friction wheels, one upon the axle and the other upon the shaft which supports one end of the motor; the latter is arranged with an eccentric operated by a lever, so that it can be thrown into gear with the friction drum upon the axle when set in motion; by this means it winds up the brake chain and checks the car.

A convenient feature of this truck is the fact that the field magnets can be slid sidewise along the two axles which support their ends, and the armature can then be slipped in the opposite direction so as to expose nearly its entire face, thus enabling small repairs to the armature to be executed quite easily without removing it. The present machine weighs little over 2,500 pounds, and is of about 30 h. p., thus replacing in power more than the two motors ordinarily used.

Winkler Gearing.—In the system shown in the cut a single 20-h. p. motor is employed per car; it is arranged to run loosely until the power is applied to the car by a friction clutch, which allows putting on the load gradually. This reduces the great initial current, and utilizes to a great extent the momentum of the armature in overcoming the static friction of the car. Both axles are driven by this single motor, the power being transmitted from the directly driven axle to the other by means of a pair of wire ropes running over pulleys of the peculiar construction shown in the cut. These give a grip on the driving rope that renders slipping less liable. The momentum of the armature is sufficient to start the car from rest even when the current is cut off before throwing on the friction clutch. The arma-

FIG. 60.—WINKLER GEARING.

ture speed is low enough to permit the use of a single reduction gear.

Goss Truck and Gearing.—In the accompanying illustration of the Goss truck, the armature shaft runs lengthways of the car, and has two gears of different diameters which mesh into two gears of corresponding different diameters fitted with friction clutches, on a shaft running parallel to the armature shaft, on each end of which are bevel gears meshing into gears on the axles. On each end of the car is placed a controlling stand with three handles. The upper one controls the speed of the car by a rheostat. The middle one is the reversing

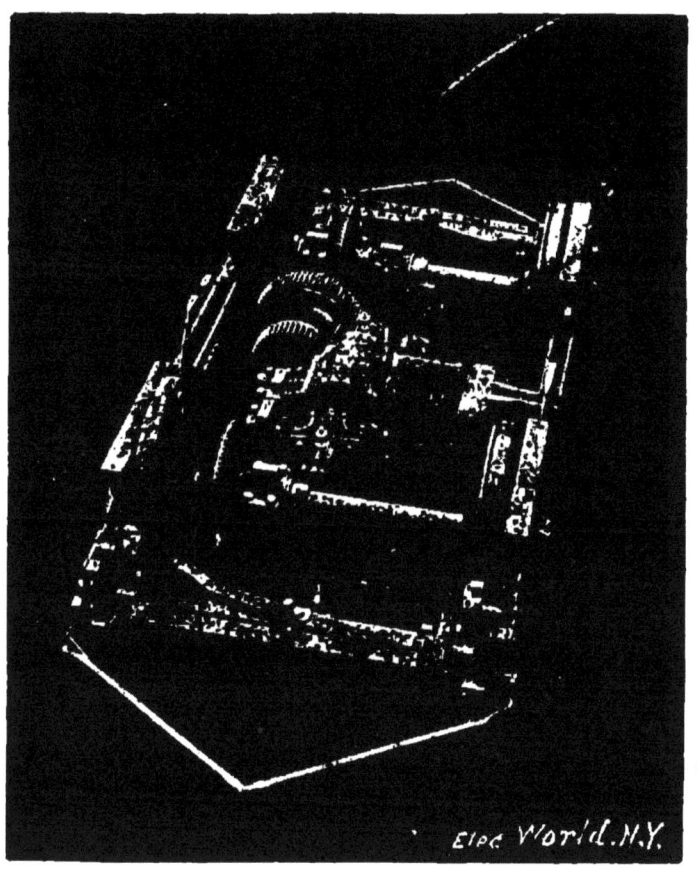

FIG. 61.—GOSS GEARING.

lever, and the lower one connects to the two clutches. By the use of this lower lever the driver has at his command by a single movement an opportunity to obtain either power or speed. On this truck with 36-inch wheels, when the lever is thrown to the left a speed of only four miles per hour is obtained, but with great power for use on hills, pulling out cars, etc. Throwing it to the right a speed of 16 miles per hour is obtained. Different speeds can be had for different roads by a change of gears on the armature shaft, etc. In one case the regular car, being partially disabled, was unable to climb the hill six miles from the power station, and called upon this car to push it up, which it did with its load, starting on the hill. The car is especially adapted for very heavy grades and sharp curves.

Loose Wheel Gearing.—In the truck shown in the accompanying illustration, the wheels, 26 inches in diameter, are loose upon the axle, and are fitted inside the hub with the Tripp roller bearing, carried on a 4 1-2 inch journal. Upon the inside of each wheel is bolted a 20-inch gear, fitting into a pinion eight inches in diameter that is keyed to each end of the armature shaft. This applies power to four different points, and gives traction upon all the wheels. The entire weight of the motor is supported by two rigid axles, thus overcoming the friction caused by the motor bearing upon a revolving axle. The truck is interchangeable and will swivel under either an open or a closed car. It will also take any radius of curve without interfering with the car sills.

Peckham Truck.—Fig. 63 shows the radial geared cantilever truck as adapted for the usual styles of motor. Figs. 64 and 65 show the yoke and journal box. The object is to obtain the maximum flexibility

FIG. 62.—TRIPP LOOSE WHEEL GEARING.

in a four-wheel truck, and to this end the axles are cushioned and are given the advantage of a perceptible though small radial flexibility. The side frames are composed of wrought iron and steel bars riveted to yoke pieces of malleable iron so as to form a practically continuous structure. The extensions that support the end springs to prevent swaying of the car body are rendered rigid by a cantilever truss, as shown in the figure. The bend in the side bar is for the purpose of allowing the free removal of the armatures sidewise. One of the most important features is the radial gear intended to give flexibility to the axle connections and to

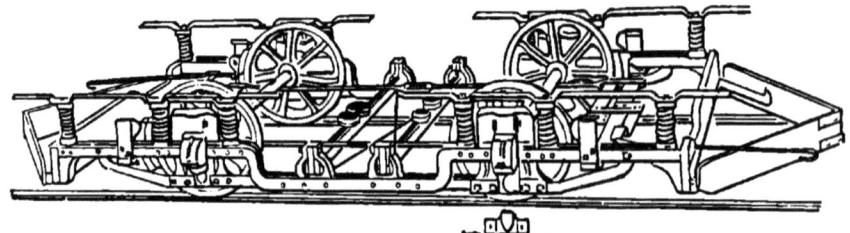

FIG. 63.—PECKHAM TRUCK.

provide cushioned supports for the boxes. The malleable iron yokes that support the side frames upon the journal boxes are formed with sockets in the upper sections, into which are inserted rubber cushions, shown at *B*, Fig. 64; the upper section of the spherical bearing *C*, Figs. 64 and 65 is constructed of the same size and shape as the apertures in the yoke; the lower sections fit into sockets in the journal boxes, but with a little more play than usual to give increased springiness in the whole structure. The whole weight of the side frames rests upon the rubber cushions inserted between the yoke and the upper section of the spherical bearings. The lower section, *E*, Figs. 64 and 65, serves simply to steady the boxes, and is carried directly by the repairing piece *E* in the lower portion of the yoke and held in position by the bolts shown in Fig. 65; removing these enables the side bars to be completely detached from the boxes. The car body is

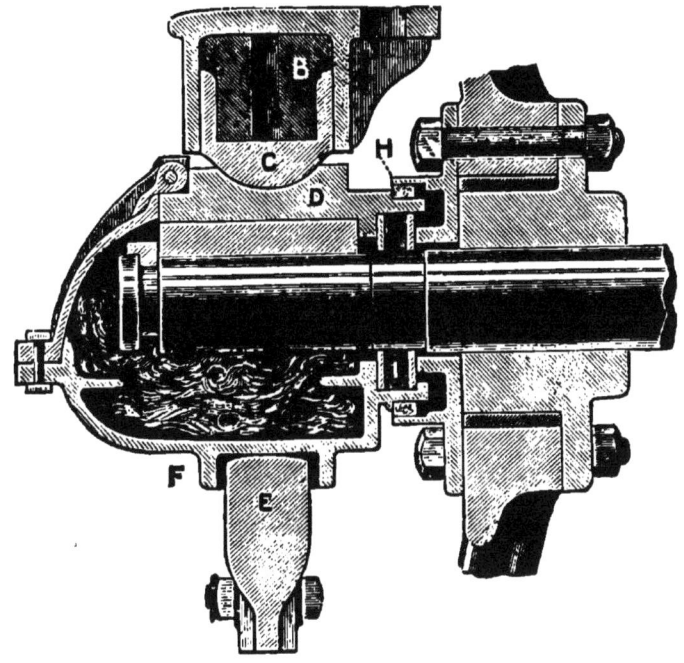

FIG. 64.

supported on the springs upon the side frames and given additional steadiness by the double plungers bearing in the yokes.

Fulton Truck.—The object in this truck, shown in the accompanying illustration, is to construct it in such a manner that the wheels are always in line and run true, preventing the uneven wearing of the wheels, as the truck is rigid.

The main sills and cross sills are made of wood, which prevents bolts working loose and rattling. A new form of brake is used, which is set on the main sills, so constructed that a direct pull is made on

FIG. 65.

each brake shoe, giving the same pressure on each wheel, and when the brake is set its position is always the same relative to the car axles, so that no effect in pulling down or lifting the car body is noticeable. If it becomes necessary to remove the wheels, the car can be jacked up and raised with

FIG. 66.—FULTON TRUCK.

the sub-sill left stationary to the body by removing four bolts on the side of the journal boxes and four others on the side of one of the main sills, and taking the lids off the oil boxes and removing the brass keys, when the whole side of the truck can be pulled off with springs and boxes attached.

Three Rivers Truck.—The special feature embodied in this truck, shown in the figure, is the principle of the equalization of strains. A cross equalizing bar is placed across one end of the truck, by means of which a three point suspension of the car body is effected, a double spring being placed in front of and a single spring at the side in rear of the front wheels. Springs are also supported on the bar at each end of the rear axle. The wheels are ground to a perfect equality in circumference, after being pressed upon the axle, thus insuring wheels that will run true with the axle. Self-oiling boxes containing simple device holding a strip of wool felt close to the axle require attention only once in two or three months, even when the traffic is heavy. The operating mechanism of the brakes is simple and entirely out of the way of the motor, while the shoe brakes are both flange and tread, and when worn can be quickly replaced by simply removing a key. The entire truck may be lifted from either or both axles by simply removing a bolt under the end of the axle.

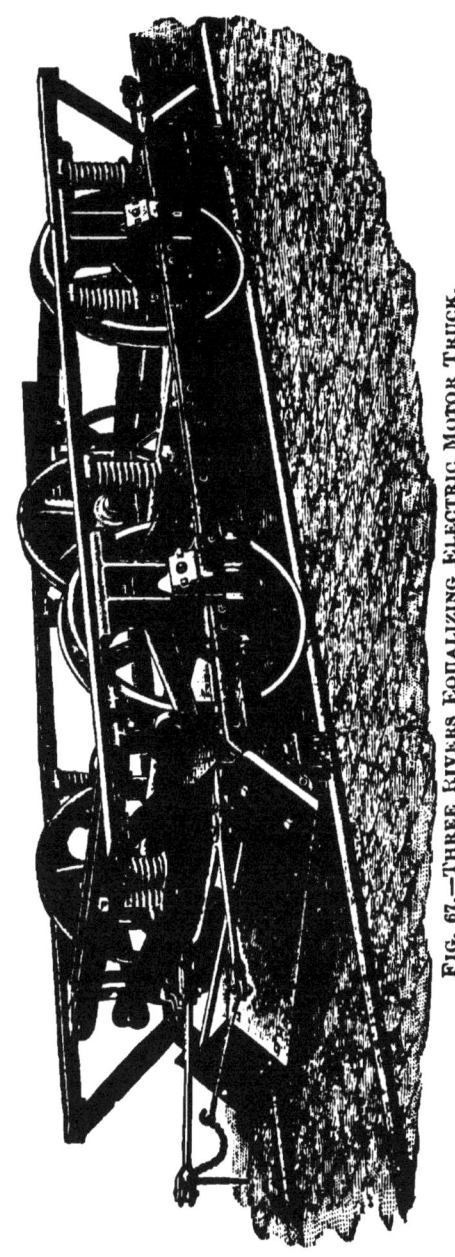

FIG. 67.—THREE RIVERS EQUALIZING ELECTRIC MOTOR TRUCK.

CHAPTER XII.

ACCESSORIES.

The following is a collection of abstracts made from some of the more important of the numerous descriptions of accessories published during the past year. Some of them must be looked upon as mere suggestions, while many will no doubt be superseded by improvements, which their defects will point out.

Trolley Arms and Trolleys.—The accompanying illustration shows a one spring compression trolley; by a proper adjustment of the set screws in the link any tension that may be desired can be obtained, from 1 to 20 pounds. The spring is inserted in a cylinder or casing so that it cannot get out of place or out of working order. It is light and is easy and convenient to handle. The top of the base stands only seven inches above the top of the car roof and the pole may be drawn down at either end of the car. The wheel is made from the best bronze, provided with hardened bearings and will run from one to two weeks with only one oiling.

The construction of the "Common Sense" arm will be seen from the adjoining figure. The arm can be used at angles varying from two to fifty degrees on either side of the perpendicular. The advantage of this trolley arm is that, as the pole is bent further away from the perpendicular, or say 40 degrees (the usual working angle), the upward pressure of the arm, instead of growing stronger as in most other trolley carriers, becomes less, thus allowing of working under bridges and other places where the trolley wire is low.

In the Lieb trolley, illustrated on page 356, the alu-

FIG. 68.—DUGGAN TROLLEY AND ARM.

minium wheel which was at first used, being found unsuitable for the purpose, has been replaced by a similar one of other material. In the trolley base the pull on the side springs is applied in such a way as to leave only a very slight resistance to the movement of the trolley pole and wheel in a lateral direction. It is said that this is the only trolley wheel made in which the spindle revolves with the wheel. The wear due to the friction will be such

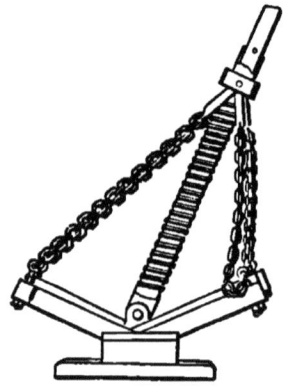

FIG. 69.—THE COMMON SENSE TROLLEY ARM.

that the trolley spindle and the bearing in which it revolves will always have a perfect fit, and the wheel will turn easily upon its centre. It is often the case when the trolley wheel turns upon the spindle that the bearing inside the wheel soon wears into a shape other than circular, causing a very

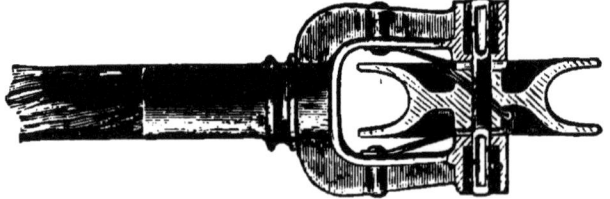

FIG. 70.—LIEB TROLLEY WHEEL.

uneven and undesirable motion of the wheel. The spindle is bored out at each end, the cavities thus formed being filled with oiled felt; three small holes extending from the cavities to the surface of the spindle serve to carry the necessary amount of lubrication to the wearing surfaces. Lignum vitæ bushings are used, and these can be replaced without removing the trolley wheel. The tension brought upon the trolley pole by the springs of the base is so evenly distributed for the different positions of the pole that it is possible to pass under a

FIG. 71.—ELECTRICAL SUPPLY CO.'S HANGER.

bridge only five inches higher than the top of the car.

Trolley Wire, Hangers, Clamps and Insulators.— The Electrical Supply Co.'s hanger shown here dispenses with soldered ears and riveted clips in sustaining trolley wires. By tightening the bolt at the top of the clamp the wire is firmly clasped, while by loosening the bolt the wire is instantly released with great ease. The time and trouble of shifting its position on the trolley wire is so small that it is easy to adjust it to any contracting or slackening of the lines. The hanger is sprung upon the span wire in the ordinary way, the tension of the wire holding it in place. The clamp screws into a plate dovetailed in the vulcanized rubber, incased in the metallic shell. It will be observed the clamps do not quite reach the lower surface of the wire and

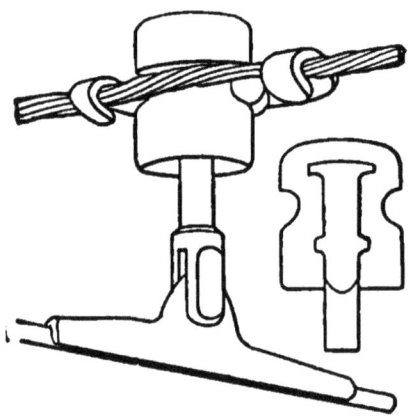

FIG. 72.—CLINCH HANGER.

are trimmed and rounded so that they escape the trolley.

In the "Clinch" hanger shown here the insulating material is of rubber, similar to that ordinarily used for combs; the stem and yoke of the insulator are of malleable iron and the pin is made of brass. Whatever strains are brought upon the insulator are entirely those of compression. There are no screw threads either upon the pin or the insulating material.

The Pierce hanger is made of malleable iron, of the bell type, and has a cavity in its interior, in which is placed a porcelain insulator of the ordinary type. This is held in position by an insulating compound, which is poured around the porcelain while hot, and filling the grooves in the sides of the bell chamber and on the side of the porcelain, holds the latter firmly in position.

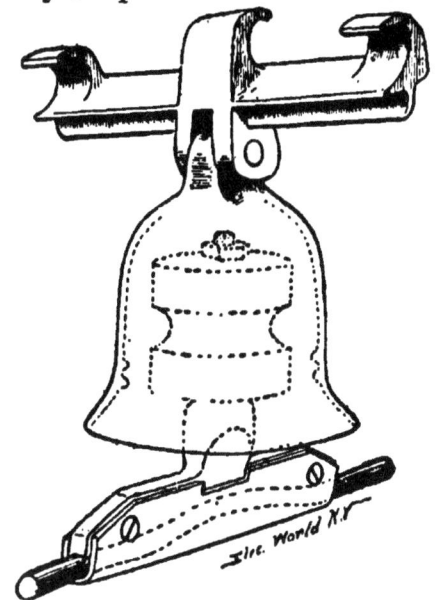

FIG. 73.—PIERCE HANGER.

The Creaghead hanger shown in the cuts was described as follows: The superiority of glass as an insulator is recognized by all. The glass insulator, screwed on the wood pin, has been used in electrical line work for years, and is recognized as the best practice for line insulation. The wood acting as a cushion adapts itself to any condition arising from unequal expansion of the wood and glass, thus securing the glass insulator against breakage. As shown in Fig. 74, this hanger consists of a grooved glass insulator into which is

FIG. 74.—CREAGHEAD HANGER.

screwed a wooden plug. The bolt for securing the clamp to the wooden plug, has a round head at the top and a thread at the bottom and is screwed into

FIG. 75.—CREAGHEAD HANGER.

the clamp. The yoke shown in Fig. 75 fits loosely in the groove of the insulator, as shown in Fig. 74. The inner curve of the yoke, coming in contact with the insulator, is a half circle with a slightly larger radius than that of the smallest circle of the groove. It is necessary to remove the yoke from the insulator about three-quarters of an inch before it becomes disengaged from the groove. The yoke can be put on and taken off of the span wire easily and without tools. All strains on the insulator tend to increase the hold of the yoke and span wire on the insulation. While the attachment of yoke and

FIG. 76.—LIEB HANGER.

trolley wire is secure, it is at the same time flexible, and will adjust itself to unequal expansion of materials.

The accompanying illustration, Fig. 76, shows a complete Lieb hanger, and Fig. 77 shows the method of construction. In Fig. 77 the metal socket in which the clip screw fits is part of a ring imbedded in the composition, and the rivet which holds the iron hood to the insulator passes through this ring, being separated, however, from it by the insulating composition. The advantage of this is that it prevents the insulator from dropping the trolley wire, no matter what accident may happen to it. Repeated blows of the trolley can break the composition, and even make connection between the span wire and the trolley wire, but the trolley wire is always supported by means of the rivet passing through the

FIG. 77.—LIEB HANGER.

ring, so that there is no danger of the trolley wire falling in the street. The clip has two jaws hinged so as to clasp the trolley wire, and these jaws are forced together by means of a taper point on the screw holding the clip to the insulator, and which forces the jaws on the trolley wire. The clip is adjustable, and will fit any size of wire, from a number zero to a number four. In attaching, the insulator is first screwed to the trolley wire and then snapped in place on the span wire. By this construction the jaws of the clamp cannot become accidentally loose, since the clamp can be opened only by turning the insulator, which involves disconnecting it from the span wire.

In the Robinson hanger shown in the figure the

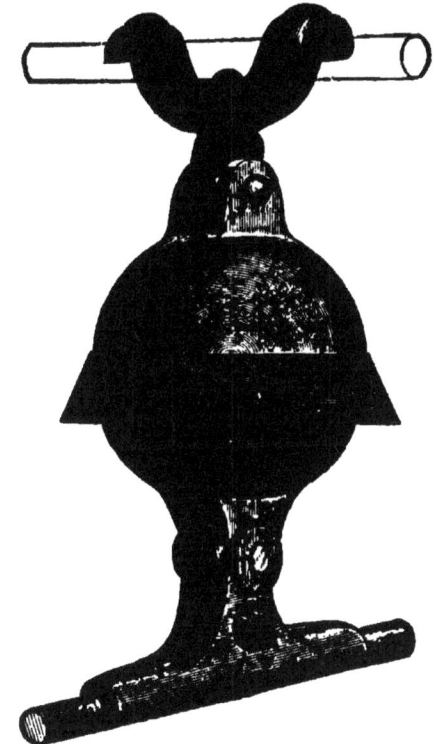

FIG. 78.—ROBINSON HANGER.

span wire clamp is of brass and can be made to clasp the span wire with sufficient tightness to prevent any slipping or side movement. The globe part is of cast iron, the two halves being separated by about one-half inch of hard rubber extending be-

yond the edge, and so shaped as to make an um-
brella-like watershed. The clamp holding the trolley wire is set up by a screw, and leaves the under
part of the wire exposed.

The Gustin clamp is formed in three distinct
parts, two sections with opposing lips, one of which
is provided with a threaded projection upon which
the clamping nut is screwed; the other with a dog
inclosed by the protruding lower rim of the nut.
Through the lower portion of these opposing sec-

FIG. 79.—GUSTIN CLAMP.

tions a groove is formed to contain the trolley wire
of any size. The opposite lips are cut at such an
angle that they still engage each other when the
largest trolley wire is inserted in the groove. The
screwing down of the nut upon the threaded projection and over the dog causes the lips to slide one
within the other, thus forcing together the lower
sections of the clamp upon the wire, preventing all
possibility of falling. The movement of the wire
from expansion and contraction is provided for in

the hinge joint included in the nut. The clamp is independent of the style of insulator used in connection with it.

The McTighe clamp is composed of a single piece of hard rolled sheet copper stamped out in such form that it may be folded up from beneath around the trolley wire, and the folded bends clamped together and supported by any of the usual forms of insulators. The final tightening on the wire is done by means of the key bolt shown in the cut,

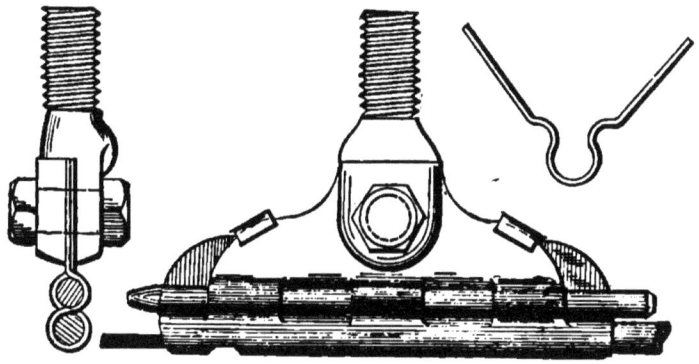

FIG. 80.—McTIGHE CLAMP.

which, acting on the interlaced portions of the clip, causes the latter to tighten around the wire like a strap. By suitable selection of the size of key, any degree of grip may be obtained, and any of the usual sizes of trolley wire may be used without changing the construction of the clip. One advantage lies in the fact that the clips may be attached to the trolley wire and to the line hangers without actually tightening the clips themselves on the trolley wire, and this operation may therefore be deferred until such times as the entire line is ready for

drawing in and straightening, and then it is necessary only to apply key bolts. After a line has become stretched and needs tightening, the hanger

FIG. 81.—STERLING CLAMP.

permits the key bolts to be withdrawn from any
desired section of the line so that the latter may be
adjusted. During this operation there is no possi-

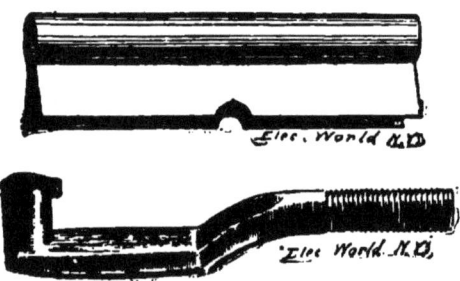

FIGS. 82 AND 83.—STERLING CLAMP.

bility of the trolley wire falling to the street, and
the motor traffic is not obstructed. For curve
hangers the suspension bolt is made with a broad

downward extension which fits the face of the clip and gives it all necessary lateral support.

The Sterling clamp is shown in the adjoining figures. Fig. 81 shows the completed hanger. Fig. 82 the clasp which passes around the wire and into the edge of a slotted block and Fig. 83 is the support, which is passed through a hole in the slotted block and is screwed into the core of the bell-shaped insulator. It will be seen by an examination of these parts that when put together and fastened into the

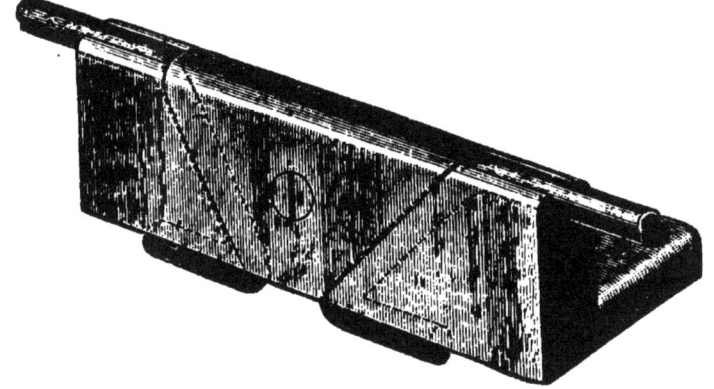

FIG. 84.—WHEELER CONNECTOR.

position which is shown in Fig. 81 they cannot be detached from each other.

The clip for connecting two trolley wires, shown in Fig. 84 explains itself. Both wires are firmly held in place with their ends bent under the clip, by a key that is held in place by a screw.

In the accompanying illustrations Fig. 86 shows a centre curve insulator made of locust wood with caps of mild steel pressed upon the tapered ends of the wooden centre piece. The eyes and caps are in one piece, making the insulator simpler and saving the expense of extra eye bolts.

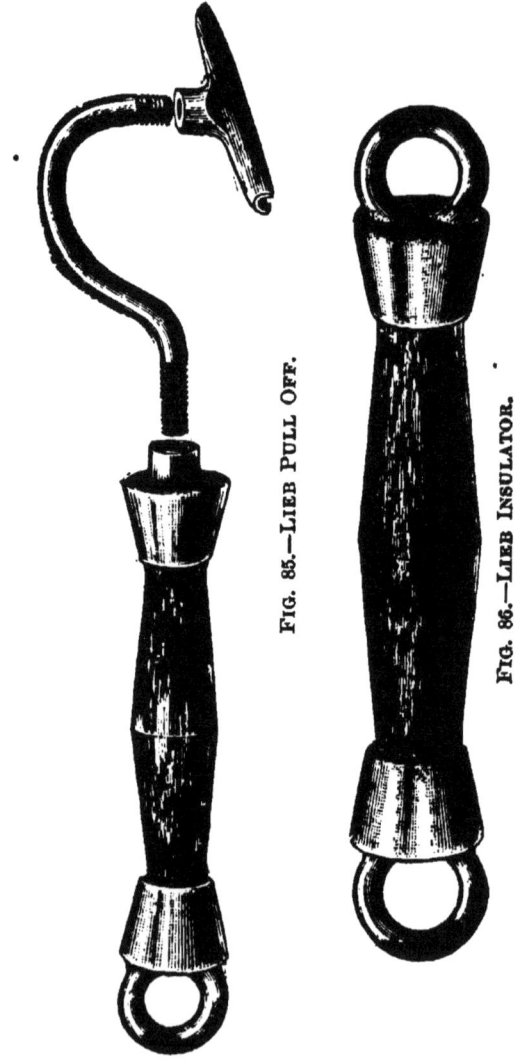

FIG. 85.—LIEB PULL OFF.

FIG. 86.—LIEB INSULATOR.

Fig. 85 shows a complete pull-off bracket, the body of which is also made of locust wood. After several years of experimenting in work, Mr. Lieb has concluded that locust wood is the best insulating material to use for a fixture that is subject to all the abuses of an insulator on a street railway curve. The shrinking or expanding of the wood in dry or wet weather does not affect the insulator, as the wood is larger at the bottom of the caps. When a strain is put on the insulator it acts as an inverted wedge, and the wood inside of the cups is compressed, preventing splitting, etc. The more strain is placed upon the insulator the tighter will it be made to fit in the caps.

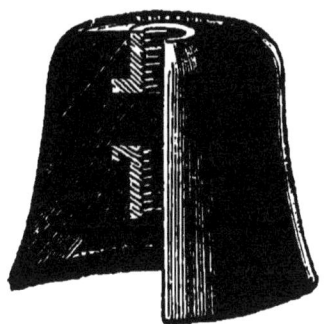

FIG. 87.—ANDERSON INSULATOR.

The accompanying figure shows a half sectional view of the Anderson railway bell insulator, made of solid insulating material. Besides possessing high insulating properties this material is claimed to be very strong, tough and durable, and absolutely impervious to water and unaffected by atmospheric conditions.

The Robinson centre curve insulator in the adjoining figure is chiefly composed of cast iron, the insulation being effected by the introduction of hard rubber bushings of ample extent and thickness to prevent leakage. He claims that the chief objection to some of the other styles of curve insulators is the lack of sufficient strength to hold a heavy line, especially in winter, when the strain of the ice is considerable.

The accompanying illustration shows the Gould & Watson molded mica insulator for span wires, which possesses the advantages of a span insulator and turn buckle combined. In setting up a span the wire is cut off to length and attached to the eyebolts, of which one is unscrewed from its case. One

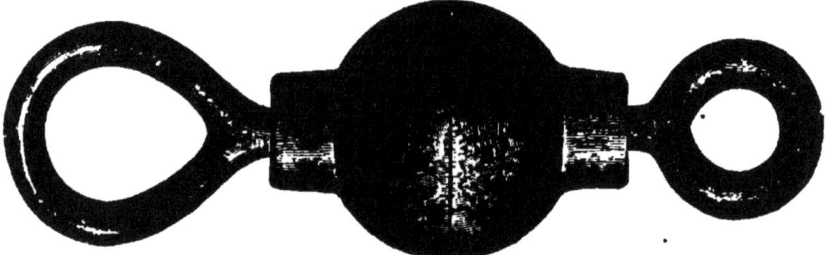

FIG. 88.—ROBINSON INSULATOR.

end of the span is hung up with the insulator on the other pole. The free end of the span wire, which is attached to the eye bolt, is strained up with a tackle until it will enter the insulator, which can then be turned up with a wrench sufficiently to strain the span wire. The strength of these insulators is said to be sufficient to break a No. 2 B. W. G. span wire without injury to the material, and it is claimed that spans can be wired with this insulator in one-half the time usually taken.

The features of the Winton insulator are the interlinking hooks inclosed or embedded within the insulating material, preferably hard rubber, the hooks being separated from each other by means of a layer of hard rubber secured between them and also covering the whole body portion, the whole being vulcanized together in one piece. Each interlocking hook is provided with a threaded stud projecting from its extremities, one of which is adapted to receive a clamping ear and the other adapted to receive interchangeable supporting pieces arranged for attachment to a supporting wire or bracket, as the case may be. The principle of the interlinking hooks gives the necessary strength to the body or bell portion, and at the same time it requires only

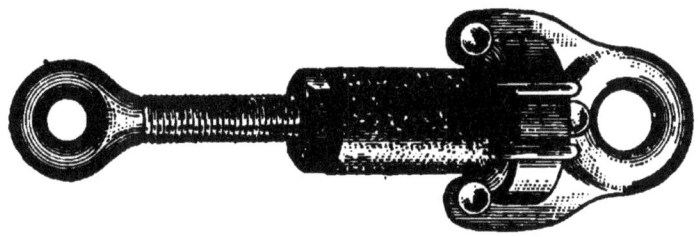

FIG. 69.—SPAN WIRE INSULATOR.

a small amount of a high grade insulating material to make a perfect insulator. For further description accompanied by an illustration see *The Electrical World*, October 24, 1891, p. 310.

Cross-Overs and Switches.—Among the cross-overs and switches described may be mentioned the following: The accompanying illustration shows an adjustable cross-over which may be adapted to crossings at any angle, and furnishes a smooth passage and unbroken contact for the trolley, whether the trolley wheel be deep or shallow. This is accomplished by making an upper flange or plate

on each of the four arms, as shown in the cut, extending a short distance out from the centre at

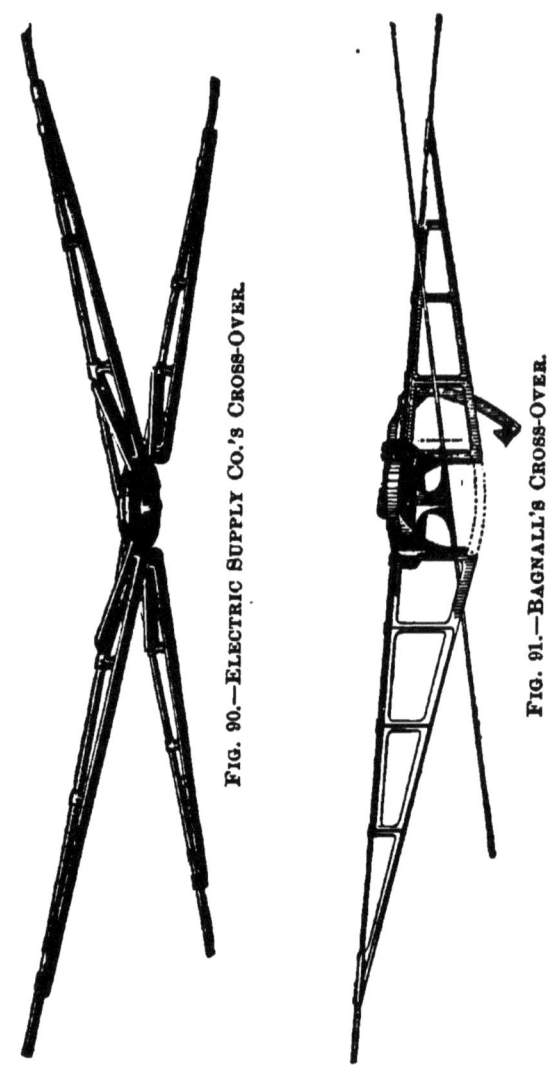

FIG. 90.—ELECTRIC SUPPLY CO.'S CROSS-OVER.

FIG. 91.—BAGNALL'S CROSS-OVER.

a slightly greater incline than that of the trolley
rib. The trolley rib is then made very shallow and

the centre post very small and short, so as to form no obstruction to the shallow wheels, the deeper wheels being provided for by the flange on the wheel coming gradually in contact with the face of the plate on the arm, which, in connection with the centre plate and that on the opposite arm, forms a smooth surface over which the trolley wheel passes. There is thus no arcing or jar in crossing, whatever the depth of the wheel may be, the flange of the deeper wheels merely reaching the plate on the arm sooner than the flanges of the shallower wheels. This is an important feature, as the depths of the wheels are continually varying by wear.

In many forms of cross-overs one of the wires is so arranged that its contact with the trolley wheel is broken. When the car passes such a crossing the lights will be momentarily extinguished and the current must be cut off from the motors to avoid an undesirable flash between the trolley wheel and wire when the break occurs. In the trolley shown in the accompanying illustration neither line is cut in placing it on the lines. It is a live crossing and cars can be run over it at full speed with the current turned on or off. The tongue is of steel and very strong, and of a peculiar shape, so that when it falls back it is well drawn up out of the way, and as the trolley wheel strikes it so far from the pin on which it is pivoted it has quite a leverage, and moves very easily. The tongue is brought back by gravity, and works with a positive motion. They are furnished, if necessary, with a sleet proof cap, so that snow and ice cannot interfere with the working of the tongue. An objection to it is that on the lower line the trolley can move in only one direction.

In the inverted switch, shown in the adjoining cut, the trolley wheel advancing on either of the diverging lines automatically throws the hinged portion of the switch over to the end of that line by an arrangement of the arms, such that the flange of the wheel striking the arm brings the movable portion of the switch into the required position. The switch is made of only two pieces, is entirely of metal, and the line passes over its back without the necessity of any cutting whatever.

Span Wire Tighteners.—The following cuts will

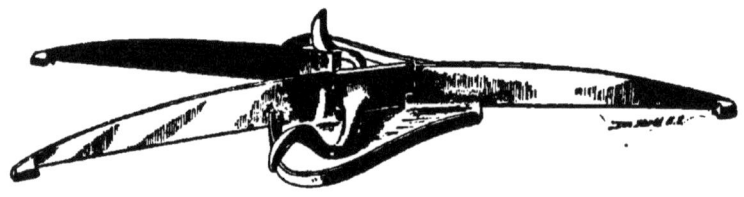

FIG. 92.—PIERCE SWITCH.

explain themselves: They can be fastened in the old way, with bolt, washer and nut, as shown, or by discarding the bolt entirely and passing the span wire through the pole to the rachet fastened at the further side. This latter method would appear to offer decided advantages.

Brackets.—The arm of the Giles bracket, shown in Fig. 95, is made of a steel tube, the truss rods being also of steel. In the construction of the bracket there are no castings except at the extreme end, where a malleable casting weighing about three-fourths of a pound is used. The upper truss rods pass on either side the pole, thus preventing any swaying motion, while the lower rod pre-

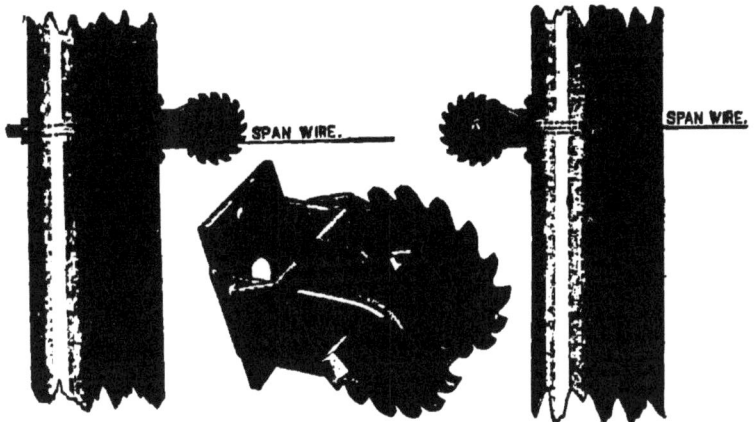

FIG. 93. SPAN WIRE TIGHTENER.

vents the bracket from being pressed upward when the trolley wheel passes under it.

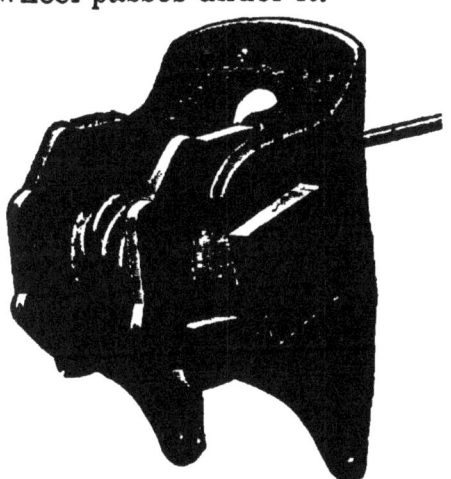

FIG. 94.—SPAN WIRE TIGHTENER.

The Duggan adjustable trolley wire bracket, shown in the cut, is adapted to be adjusted to the supporting pole whether the pole be straight or crooked. It commends itself when perfectly straight poles are expensive or hard to obtain.

Miscellaneous.

Pullman Car.—As seen in the accompanying illustration, this car has two decks, and is designed to double the carrying capacity of street railways,

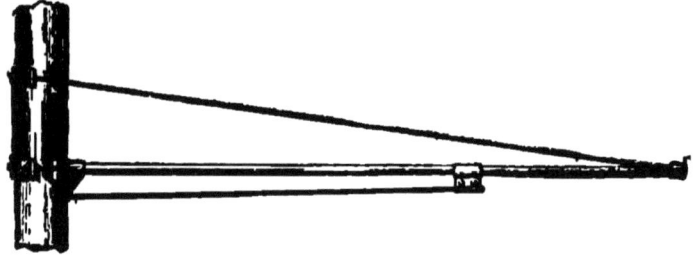

FIG. 95.—GILES BRACKET.

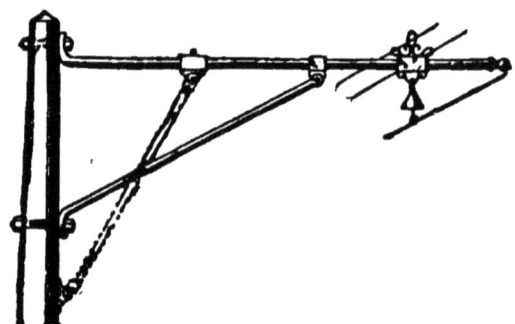

FIG. 96.—DUGGAN BRACKET.

as well as adding to the comfort of passengers. It is 32 feet long, 7 feet and 4 inches wide, with a height of 14 feet 9 1-2 inches. A seating capacity for 40 passengers on each deck is provided for, the car body being arranged so that passengers may enter at the centre of either side, spiral stairways leading to the upper deck. Two compartments comprise the lower car body, each 12 feet long, with circular ends, the seats being arranged around the ends and sides. A canopy covers the top. It weighs 28,000 pounds and cost $3,500. The car exhibited was operated by the trolley system, Westinghouse motors being used, of 25 horse power. It is equipped with electric chandeliers and electric heaters in each compartment, the finish being in

FIG. 97.—THE PULLMAN CENTRE VESTIBULE CAR.

mahogany, with a handsomely painted exterior. It rests on two trucks of special design, having double brake attachments and a friction brake. The interior conveniences introduced are electric signal bells for stopping and an electric diagram showing vacant seats. The services of three employés are necessary in its operation. It is thought that it will be adopted by roads where traffic is exceedingly heavy, as each car will carry 250 passengers.

Snow Sweeper.—In the McLain track sweeper the various parts are so arranged that the running gear and propelling motors may be utilized for the transportation of passengers during the summer season and for a sweeper in the winter. The main object had in view was to overcome the difficulty due to the defective transmission of the motive power whenever the rotary brushes or track sweepers were necessarily raised or lowered in the operation of the machine, and to afford ready and convenient means for accompliishing this part of the operation. Two 15 h. p. motors are mounted upon a truck of ordinary construction in the usual manner, being geared with the axles of the traction wheels and regulated by a rheostat operated from either end of the machine. The platform of the car is rhomboidal in form, so that the two rotary sweepers located at each end of the machine and placed obliquely to the track may be raised to a point even with or slightly above the edge of the platform if desired. Upon the platform are secured two 15 h.p. Thomson-Houston motors which operate their respective brushes independently, each being regulated by a separate rheostat. The axial shafts of these motors are geared with their countershafts in the usual manner, the only change in the form of gearing being in the substitution of sprocket wheels, belted with sprocket wheels on the axles of the sweeping brushes, for the toothed pinion wheel ordinarily gearing with the toothed wheels of the car axles. The sweeping brushes are of steel wire

formed in four sections, bound together, the sets of
steel wire projecting radially at right angles relative
to each other.   The device for raising and lowering
the sweeping brushes consists of two screw-
threaded vertical shafts located on each side of the
platform, one of which is provided with a hand
wheel, and each having a sprocket wheel, about
which a sprocket chain passes, so that both of the
shafts may be operated simultaneously from one
point.

These horizontal levers being pivoted on a line
with the countershaft carrying the driving sprocket
wheels permit the raising and lowering of the
sweeping brush without in any wise slackening the
sprocket chain about the sprocket wheel on the
countershaft and the sprocket wheel of the brush,
thus always preserving the transmitting chain taut
whatever may be the position of the brush.   From
the centre of the platform a vertical mast rises to
which the trolley pole is pivoted, a cluster of lamps
being disposed in a circle at its upper end.   Two
horizontal levers are pivoted on a line with the
countershaft carrying the driving sprocket wheel,
threaded wrought iron blocks being provided.
located under each lever through which the vertical
shafts pass.   These levers extending forward are
bent downwardly over the edge of the platform
and constitute the pivots of the brush.   For an illustration see *The Electrical World*, January 17, 1891,
p. 43.

Snow Plow.—A snow plow of the Eastern Electrical Supply Company was illustrated in *The Electrical World*, October 24, 1891, p. 309, and is designed

with foundations to accommodate any motor system. A V-shaped wooden nose or plow reaching over the track 15 inches is attached to each end, which serves to throw the snow away on each side, while a leveling wing extends out 4 feet, hung fore and aft, and held in place with hooks. These wings are raised with ropes and stand up at the side of the body when not in use. To clean the track of snow and ice four diggers are applied, made of 2 1-2 inch square Norway iron, with steel feet, which are operated by levers on the body of the plow, with four grounding brushes.

Snow Cleaner of the Thomson-Houston Co.—Mr. Barr of that company described their new cleaner as follows: "The machine we had last year was a brush of rattan set at an angle of 60 degrees, revolving in front of a frame, and driven by a motor by means of sprocket chains. In a light sugar snow that brush gave good satisfaction; but in a heavy wet snow the brush would clog, and served rather to pack the snow on the rail and make it solid; and, therefore, rendered it impossible for the wheel to get traction. The sprocket chain is always a source of trouble and annoyance. The lines were laid down to follow, first, to do away with the rattan broom, and second, to get a positive drive for the cylinder. We have built a broom with steel teeth. It is supported from 36-inch 400-pound wheels, on a '22a' Bemis box. The wheels are driven in the same way as with cars, the motors being placed on the axles. We are using single reduction motors for them. The cylinder is set at an angle of forty-five degrees to the track extending completely across,

and is hung from a rocker shaft. The motor coun-
terbalances the blade of the broom around that
shaft, and is geared directly through spur gearing
to the flyer. The main feature of the flyer is that it
has a series of blades. The best description of it is
to refer you to the paddle-wheel of a steamer only
of smaller diameter, having the blade cut at the
centre, to allow the spur gearing for the drivings.
These blades are of steel plate, about a quarter of
an inch thick, and there are eight of them on each
flyer. To the back of the blades are bolted steel
brushes, the brushes being made of flat steel wire,
cutting edgeways. These brushes are adjustable,
and ordinarily their surfaces project from five-
eights to one and one-half inches beyond the blade.
The blade does the major part of the work. It
breaks the heavy snow, and will actually cut ice.
The steel brush does the rest of the work. It sweeps
the road clean, and if allowed to remain long
enough in one place, will cut the ice as well. The
motor driving the broom or brush is independent of
the motors driving the car, so that the sweeper can
be propelled in a light snow at a speed sufficient to
keep it ahead of any of the cars on the line. If
heavy snow or ice should be encountered on the
track, the sweeper can be slowed down, the brush
steel keeping its normal speed. By doing this, these
brushes, made of steel or wire, will actually cut the
ice clean down to the rail. The whole thing is very
novel, but it has done good work. We have experimented some with a new machine, somewhat
different from the old-one. Where the latter only
had four blades the new one has eight; the old

one was set at 60 degrees angle, the new one at 45;
in the old we used the sprocket chain, in the pres-

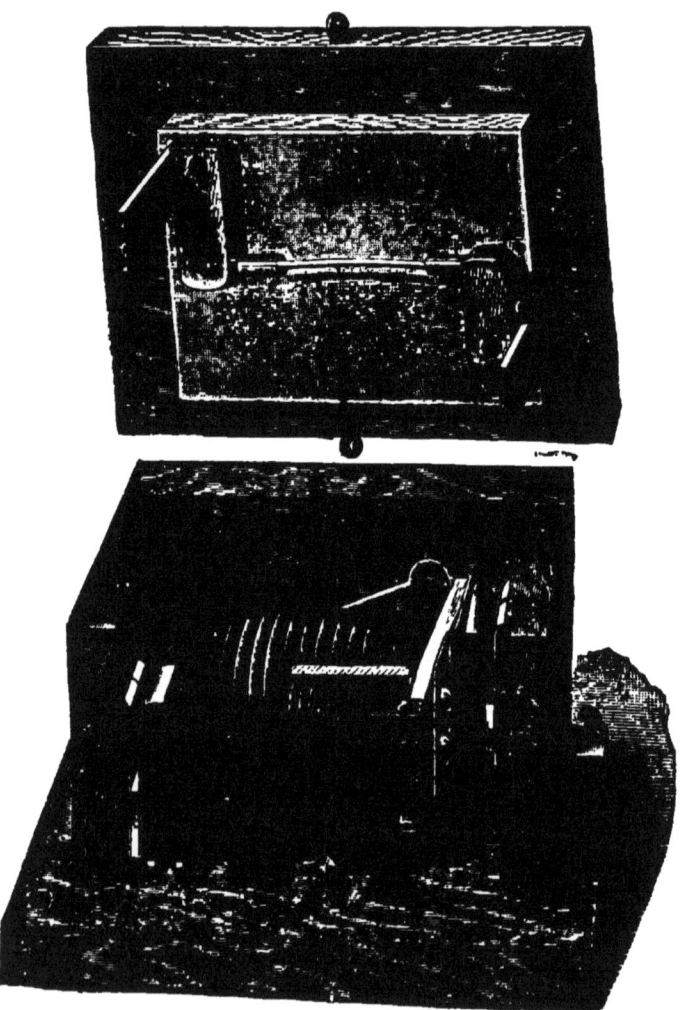

FIG. 98.—HARRINGTON CUT-OUT.

ent the spur gearing. We had a 10-h. p. motor but
now have a 25. In the old one the blade made 80

turns a minute, in the new one it will run up to 150 turns a minute."

Magnetic Cut-Out.—As a simple fuse is not to be relied on in all cases, the device shown here is made by combining a magnet with a fuse. A properly proportioned magnet is placed in shunt with a fuse which can carry but a small part of the maximum safe current. The resistance of the magnet being low, the fuse is practically shunted out until the current through the magnet becomes too great. Then the armature is drawn up, and the whole current is sent through the small fuse, which is sure to blow. A spring catch holds the armature until the fuse is replaced, which can be quickly done without the use of any tools. The magnet is protected from the molten metal by a thick sheet of asbestos.

Car Lightning Arrester.—The one shown in the accompanying illustration, consists of two air chambers, having vents through which a curved carbon, hinged at its centre of curvature can freely pass. The carbon point in the chamber should be set so that it is separated from the curved carbon when in its normal position by one-sixteenth of an inch. The distance between the two carbons should likewise be one-sixteenth of an inch when the curved carbon is in the other chamber. A lead fuse, and an insulating block, with a slot through which the lead fuse is passed, are provided. Connections to the motor and trolley are made as shown in the figure. The action of the arrester is as follows: When a discharge takes place it passes through the lead fuse to one of the carbons, then across the air space to the curved carbon and from there to the ground, as shown in the figure. The dynamo current then following causes an arc to be established between the carbons. The heat generated by this arc expands the air in the chamber, increasing the air pressure, and causes the curved carbon to be instantly blown from one chamber to the other. This ruptures the arc and adjusts the arrester for the next discharge. In the reported tests upon this arrester, 500-volt and 1,000-volt gen-

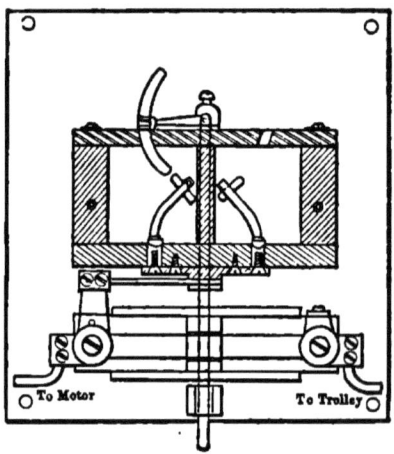

FIG. 99.—WESTINGHOUSE CAR LIGHTNING ARRESTER.

erators have been repeatedly short-circuited through it. In every case the circuit has been instantly broken without injury to the dynamo, and the arrester has as quickly set itself in readiness for future use. In one of the tests made to demonstrate its promptness and reliability, the carbons in the chambers were so adjusted as to touch the curved carbon in either of its two normal positions. A 1,000-volt generator was then short-circuited through the arrester, which resulted in the circuit being several times automatically opened and closed in one second without injury to either dynamo or arrester.

Safety Devices for Electric Wires.—T h e Electric Wire Safety Attachment Company's safety device shown in the adjoining cut has for its object the rendering harmless of a broken line wire. The plate to which the end of the wire is attached is inserted in the slide at the end of the wire projecting from the tie or supporting wire, which is slotted to receive it, thus forming the connection and holding it in position unless the wire should break. In this case the springs would draw this plate out and away from the other, allowing the trolley wire to drop down, thus breaking the connection and rendering the broken wire per-

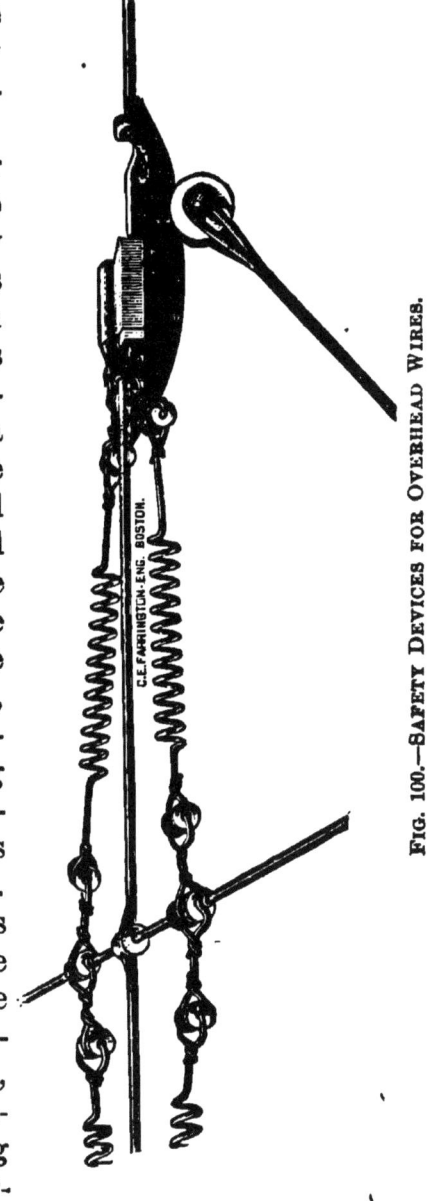

FIG. 100.—SAFETY DEVICES FOR OVERHEAD WIRES.

fectly harmless until it could be repaired. An advantage claimed is the facility of attachment, as the springs may be put in place and removed if necessary without interfering in any way with the line.

Sand Distributor.—The trouble which has been experienced in the ordinary forms of distributors is the clogging of the sand in the narrow neck of

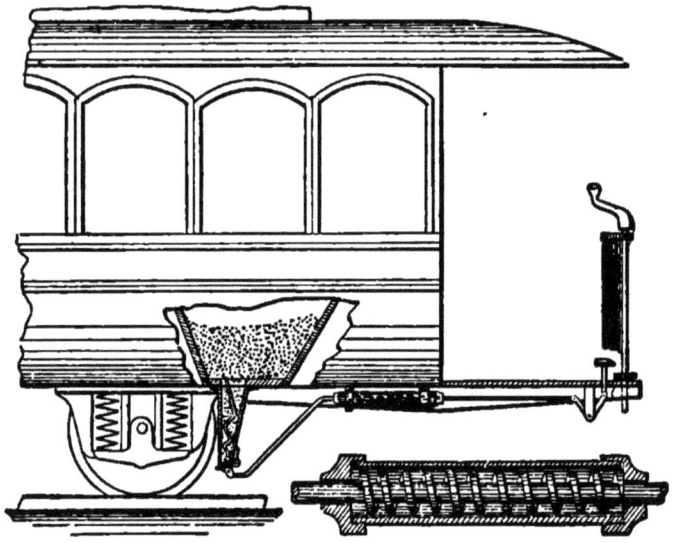

FIG. 101.—BATES SAND DISTRIBUTOR.

the containing bag. To overcome this, a corrugated metallic ribbon is in the form shown made to extend from the opening, up through the sand, the slightest jar being sufficient to keep the sand from becoming fast in the neck.

Speed Recorder.—The principal parts of this instrument are a rotary pump, a cylinder and piston. Oil is used as a circulating medium, the pump

chambers and cylinders being entirely filled. While
the machine is at rest, the piston, to which the
gauge-wire and pencil are attached, is retained in
its lowest position by two coil springs; but when
given motion the pump produces a pressure of oil
beneath the piston, causing it to rise to a position
where an equilibrium is established between the

FIG. 102. —BOYER RAILWAY SPEED RECORDER.

pressure of oil and the tension of the springs; this
point is determined by the speed at which the pump
is moved, each thirty-second of an inch rise of the
piston indicating a speed of one mile per hour.
Moving around a drum in the upper part of the
machine at the rate of one-half inch to the mile is a
ribbon of paper, having horizontal and perpendicu-

lar lines, each horizontal line from the base representing five miles per hour, and each perpendicular
line a mile traveled by the car. If the car is moving
at the rate of 20 miles per hour, the pencil will trace
its mark on the fourth line from the base, and for
every mile traveled the paper will move under the
pencil one-half inch, or the exact distance from one
perpendicular line to another. By examining the
chart the exact speed at which the car passes any
point on the road, the number and location of stops,

FIG. 103.—DUGGAN'S RAIL CHAIR.

the distance, speed and location of any backward
movement that may have been made, can be determined at a glance. A gauge is provided, the needle
of which indicates the speed at any given time. The
pencil is a brass wire, and when in good condition
makes a distinct line on the prepared surface of the
paper. It is held by friction and is forced against
the paper on which the record is made by a spring
within the holder.

Rail Chair.--In the railway chair, shown in the illustration, there are no bolts, rivets or wedges used. Paving blocks may be laid close to the rail, and the rail may be removed at any time without drawing the spikes or changing the position of the chair.

Connector for Street Car Lighting Circuits.—In the accompanying illustration of the Armstrong connector for the lighting circuits of motor and town cars on electric railways, the two parts are

FIG. 104.—CONNECTOR FOR STREET CAR LIGHTING CIRCUITS.

interchangeable, and the projecting metal tips are in circuit only when the two parts are slipped together; when the parts are separated these ends are dead and may be handled with safety, even in wet weather. The end of each conducting cord is soldered into a brass tube, and a knot in the cord takes the strain off this joint. The contact surfaces are large, and a stiff spring insures a quick make and break. The insulating bushings, as well as the shell, are made of hard rubber. The whole connector is not much larger than the illustration.

IF YOU WISH TO KNOW

SECOND EDITION.

Revised.   Enlarged.   Entirely Rewritten.

HOUSTON'S DICTIONARY
OF
Electrical Words, Terms and Phrases.

562 Pages.   570 Illustrations.   Price, $5.00.

The plan of the Dictionary is such as to bring it within the range of the entire community.

To electricians and electrical engineers, and to the superintendents and others at the head of the great electrical industries of the country, it ought to prove literally worth its weight in gold as a handy book of ready reference.

To the editor or journalist; to the intelligent reader of scientific periodicals, as well as of the newspapers and magazines; to the school teacher, the college professor, the lawyer, the doctor, the professional man generally, it will be a place where he can satisfactorily find the proper meaning of electrical terms.

To students of general electricity the book will be necessary as supplying what no other dictionary in the English language has hitherto attempted to define. To students in colleges or schools it will be indispensable as a labor saver, enabling them to find in concise form the general heads under which electrical terms are arranged.

To manufacturers, contractors, engineers; to the army of electrical linemen, operatives in electrical manufactories, runners of dynamos, drivers of motors and electrical tramways, and to the hosts connected with telephony and telegraphy, both as operators and in construction work, to the numerous manufacturers and constructors of hotel annunciators, burglar alarms, electro-plating devices and the like, the book, we feel confident, will prove invaluable as enabling them to find expressed in plain and simple language the exact meaning of the terms they so frequently employ and how they should be used.

And last, but not least, in this Electrical Age the Dictionary is positively an actual necessity to the general public, to enable them intelligently to use their mother tongue, which is now becoming thoroughly imbued with electrical words, terms and phrases, with which they constantly come in contact in conversation respecting the lighting of their streets, houses and factories, the driving of their system of cars, the operation of telephone and telegraph lines, the protection of their houses by burglar alarms, the regulation of house temperature by thermostats, the adornment of their homes by electro-metallurgical process and the many other of hundreds of ways in which electricity has been made useful to man.

Some idea of the scope of the work and of the immense amount of labor involved in it may be formed when it is stated that the dictionary includes upward of 5,000 distinct words, terms or phrases. Each of the great classes or divisions of electrical investigation or utilization comes under careful and exhaustive treatment, and while close attention is given to the more settled and hackneyed phraseology of the older branches of work, the newer words and the novel departments they belong to are not less thoroughly handled. Every source of information has been referred to, and while libraries have been ransacked, the note-book of the laboratory and the catalogue of the ware-room have not been forgotten or neglected. So far has the work been carried in respect to the policy of inclusion, that the book has been brought down to date by means of an appendix, in which are placed the very newest words, as well as many whose rareness of use had consigned them to obscurity and oblivion.

The scheme of treatment is as follows :

1st. The words, terms, and phrases are invariably followed by a short, concise definition, giving the sense in which they are correctly employed.

2d. A general statement then follows of the principles of electrical science on which the definition is founded.

3d. When, from the complexity of the apparatus, or from other considerations, it has been thought desirable to do so, an illustration or diagram of the apparatus is given.

4th. To facilitate study, an elaborate system of cross references has been adopted, so that it is as easy to find the definitions, as the words, and aliases are readily detected and traced.

In applying these rules great care has been exercised to secure clearness, to the end that while the definitions and explanations shall be satisfactory to the expert electrician, they shall also be simple and intelligible to those who have had no training at all in electricity or are novices in the art. This is work of some difficulty, but Professor Houston has successfully achieved his purpose. No one will regret the detail into which he goes, but, on the contrary, in view of the fact that so many of his definitions are new, and are not to be found elsewhere, in any form, every one will be glad that the latest terms in vogue in the most recent applications come in for elaborate, yet "popular," treatment

The dictionary is a handsome book. The typography is excellent, being large and bold, and so arranged that each word catches the eye at once by standing out in sharp relief from the page. The volume is convenient in size, and the binding and paper are perfect. In a word, the mechanical production of the book has been given special attention, and no cost has been spared ; but it is placed within the means of all who have an interest in a great, new and fascinating department of modern knowledge and discovery.

What is Said of the Dictionary.

The generally-recognized necessity that has existed for a book defining electrical terms is clearly shown by the great sale the Dictionary has enjoyed from the very first, and which it still continues to enjoy, while the enthusiastic words of praise regarding it from purchasers and from newspapers and scientific journals of every class, from one end of the country to the other, and the unanimity with which they recommend it, demonstrate beyond peradventure how satisfactory the work proves on examination and, therefore, its great value to all in any way interested—and who, nowadays, is not?—in electrical progress and development.

The following opinions, selected almost at random from among hundreds received, will give an idea of the views expressed :

It is just what I need; it fills the bill.

J. H. HARDING, La Porte, Ind.

Invaluable to those interested in electricity.

A. C. TERRY, Buffalo, N. Y.

It is an excellent work and very complete, and one that ought to be in every library.

E. W. KITTREDGE, Minneapolis, Minn.

Interesting and instructive even to those only remotely interested in electrical subjects.

GEO. B. PRESCOTT, JR., New York.

Houston's Dictionary is the most valuable of any single book belonging to the literature of electricity.

L. R. CURTISS, Mendota, Ill.

Fills a very large gap that previously existed in electrical literature.   A. E. KENNELLY,
Edison Laboratory.

Involved a vast amount of labor, and the work has been well done. It ought to be of much use to a very large class of men engaged in electrical pursuits.
H. S. CARHART, Prof. of Physics, Univ. of Mich.

The first impression is one of satisfaction, and this grows with further examination. Must prove valuable both to the professional man and the general reader.   GEORGE A. HAMILTON, New York.

The book is one that every man engaged in electrical work should have on his desk or in his library. I cannot speak too highly of it, and shall certainly recommend it among my friends.
JOHN INGALLS CARMICHAEL, E. E.,
Bridgeport, Conn.

I desire to congratulate you upon the publication of an Electrical Dictionary. A reference book of this kind is of great importance, and your Dictionary is one of the most valuable contributions to electrical literature. The work shows that great care has been bestowed upon it. It is accurate, and must be of the greatest value to every one interested in electrical matters, particularly to the student of electricity.
S. S. WHEELER, Expert, Board of Elec. Control, N. Y.

It is not simply a classification of subjects and terms culled from other sources in a haphazard manner, but the entire work bears throughout the impress of rare judgment, discernment and high scientific intelligence, and the latest developments of electrical science are briefly and admirably presented. It is *multum in parvo*, and will prove an invaluable reference book to both students and professionals.
F. W. JONES, General Manager and Electrician the Postal Telegraph and Cable Co., New York.

Press Opinions.

The following are only a few of the many flattering notices of the Dictionary that have appeared, but they are fair samples :

Engineering News.

A valuable and convenient work of reference.

London Electrical Review.

The name of the author is a sufficient guarantee of the excellence of the work.

Light, Heat and Power.

Professor Houston, being an electrician of prominence, is peculiarly well fitted for his task.

Practical Electricity.

By far the most useful book ever printed on this continent for the benefit of electrical students.

Age of Steel.

In view of the wonderful expansion of the electrical industry, this new work is almost indispensable.

San Francisco Chronicle.

Virtually a condensed encyclopedia. It fills a new field, and may be warmly commended for its fullness, accuracy and good arrangement for ready reference.

Indianapolis Journal.

The book contains all the technical terms and phrases now used by electrical scientists and inventors, with definitions and applications. It might be called an electrical cyclopedia.

Philadelphia Times.

In this age of common and almost universal use of electricity, a dictionary of electrical terminology becomes a necessity. Professor Houston's work is very carefully prepared, and it will be found useful both by electricians and the general public.

THE ELECTRIC RAILWAY
IN THEORY AND PRACTICE.

By O. T. CROSBY and Dr. LOUIS BELL.

This timely book, just issued, treats all departments of the Electric Railway of the present day as comprehensively as is practicable in a volume of reasonable size.

The illustrations have been prepared especially for it, and many of them will prove entirely new to the electrical public.

The intent of the work is to place before all who are interested in the subject of electrical traction — whether electrically, financially, or simply in a general way—an explanation of the general principles and methods which have led to the model electric railway, and the latest information as to the methods that have proved and are proving successful in practice. At the beginning the rudiments of general electrical theory are explained in a manner as simple as possible, and the reader is led on to the comprehension of the principles that underlie the design, construction and operation of the electric motor, especially the form which is used for the propulsion of street cars. Having considered the motor in general, the next topic to engage the reader's attention is the ultimate source of the power which is utilized in the motor; in other words, the mechanism by which the operating machinery is driven — engines, water-wheels, and the like.

◡Having thus obtained a general view both of the electrical transmission of energy and the generation of mechanical energy from natural sources, the treatise goes on to take up more in detail the principles of electric railroading.

Then the subject of street car motors proper, considered both electrically and in their relation to the mechanical problems that have to be met, is taken up at considerable length, and a series of working instructions is given for the proper care of such apparatus. The line that furnishes power to the moving motors is the next subject for consideration, and the principles of the electrical transmission of energy—so far as the working conductors are concerned—are fully set forth, together with various details of the applications of these principles to every-day electric roads and suggestions for their proper equipment. Here, too, properly should be mentioned a complete set of rules for the erection of overhead lines, fully illustrated with diagrams and cuts of the apparatus. Following this the design, arrangement and proportioning of power stations to fulfill any required conditions is taken up, and in connection with this a considerable amount of instruction is given concerning the practical operation of the dynamo, particularly with reference to the accidents that are likely to befall it and the ways of remedying them.   ·

The general consideration of the subject is

then completed by an exhaustive chapter on the efficiency of electric traction, both with reference to what has already been accomplished and to the lines of possible improvement for the future. Next in order come chapters treating of the use of the storage battery in connection with electric traction as a separate problem, the underground conduit and its various modifications, together with telpherage and similar unfamiliar applications of electricity.

Chapter X, is one that should be especially commended to all who are interested in engineering problems having for their objective point the facilitating of rapid transit. It is a discussion as exhaustive as the state of the art permits on the application of the electric motor to regular railroad work, substituting for the ordinary locomotive a simpler, more efficient, and more manageable apparatus in the electric motor. Following this is a chapter which should be very useful to those who are operating or intend to operate electric railroads. It is a treatment of the commercial side of the problem, cost of installation and operation, the probabilities of a paying traffic, and the financial aspect of the question generally.

Finally, the volume closes with a brief resumé of the history of electric traction from the invention of the electric motor up to the period of development in which we are living to-day.

Street Railway News.

The work presents both the elementary theory of electrical traction and the general features of the best practice, describing in detail particular methods and forms of car machinery only so far as they are of importance in illustrating the broader principles on which they depend * * * The electric railroad at the present time is attracting more attention than any other branch of the electrical industry, and this book appears at a time when all the information possible on the subject is being sought after.

Scientific American.

With nearly 150 illustrations this book is a very good contribution to one of the most important branches of electrical engineering. What the railroad of the future will be,and what part electricity will play in its development is altogether conjectural. This book tells what the aspect of the subject is to-day. The subjects of prime motors, electric motors, and car equipments, the line track and station economy, storage battery traction, high speed service,and commercial considerations are typical subjects. In the five appendices considerable useful information is given, notably a section on lightning protection, by Professor Elihu Thomson.

Street Railway Gazette.

Every now and then there appears among the literature of every industry some work that can be considered a standard. Only once, however, comes a work that can be called *the* standard on the subject treated. Of such a nature is the work of Oscar T. Crosby and Dr. Louis Bell, entitled "The Electric Railway in Theory and Practice," published by the W. J. Johnston Company, Limited, Times Building, New York. * * * The thorough scientific knowledge of Dr. Bell is very happily blended with Mr. Crosby's equally thorough practical knowledge of the same subject. It is highly scientific without being scientific, that is, it is perfect science treated in a way to be clearly understood without further and more painstaking study.

Railroad Gazette.

This eminently practical treatise on the methods of operation and construction of electric railroads and their equipment is a real addition to the literature of the subject.

No other work gives much that is of practical use to the operating officers and the mechanics of such railroads.

The chapter on motors and equipment contains general instructions as to the care of motors and their operation which are well prepared and are a valuable addition to the work. Anyone reading this chapter can, from it, become comparatively well informed upon the purpose and construction of the mechanism which has directly to do with the starting and stopping of cars.

The cuts showing the characteristics of the winding on the different kinds of dynamos and motors are exceptionally explicit, and the meaning of the words "series" and "multiple," so confusing to the layman, are explained in such a simple way, on page 15, that no one could fail to understand them. There are some general deductions about electricity which have been drawn from complex experiments that are so complex when expressed in words that it is next to impossible for the beginner to understand them, yet in this work they are clearly shown by diagrams and curves, and so well lettered and arranged that they are easily comprehended. This is particularly true of the varying efficiences of motors under different conditions with different amounts of current and running at different speeds. So, too, with the magnetic properties of various kinds of iron and steel. The diagram for this last shows forcibly the great value of soft annealed irons for the field magnets and armatures of electric motors of light weight.

The part of this work which is of most direct interest to the capitalist and the investor in street railroad stocks is that on the efficiency of electric traction. Diagrams are given showing the efficiency of the dynamo itself under a varying load, of the plant as a whole, and of motors combined with the plant.

————

"THE ELECTRIC RAILWAY" or any other Electrical or Street Railway Book published, will be mailed to any address in the world, postage prepaid, on receipt of price.

Address

THE W. J. JOHNSTON COMPANY, LTD.,
Times Building, New York.

ELECTRICITY AND MAGNETISM

A Series of Advanced Primers,

By Prof. EDWIN J. HOUSTON,

AUTHOR OF

"A Dictionary of Electrical Words, Terms and Phrases."

Cloth. 225 pages, 128 illustrations. Price, $1.00.

Prof. Houston's Primers of Electricity written in 1884 enjoyed a wide circulation, not only in the United States, but in Europe, and for some time have been out of print. Owing to the great progress in electricity since that date the author has been led to prepare an entirely new series of primers, but of a more advanced character in consonance with the advanced general knowledge of electricity.

Electricians will find these primers of marked interest from their lucid explanations of principles, and the general public will in them find an easily read and agreeable introduction to a fascinating subject.

CONTENTS.

I.—Effects of Electric Charge. II.—Insulators and Conductors. III.—Effects of an Electric Discharge. IV.—Electric Sources. V.—Electro-receptive Devices. VI —Electric Current. VII.—Electric Units. VIII. —Electric Work and Power. IX.—Varieties of Electric Circuits. X.—Magnetism. XI.—Magnetic Induction. XII.—Theories of Magnetism. XIII.—Phenomena of the Earth's Magnetism. XIV,—Electro-Magnets. XV.—Electrostatic Induction. XVI.—Frictional and Influence Machines. XVII.—Atmospheric Electricity. XVIII.—Voltaic Cells. XIX.—Review, Primer of Primers.

PUBLISHED AND FOR SALE BY

THE W. J. JOHNSTON COMPANY, Ltd.,

TIMES BUILDING, NEW YORK.

ELEMENTS OF

STATIC ELECTRICITY,

'WITH FULL DESCRIPTION OF THE HOLTZ AND TÖPLER
MACHINES AND THEIR MODE OF OPERATION.

By PHILIP ATKINSON, A.M., Ph.D.

Cloth, 12mo; 228 Pages; 64 Illustrations.

PRICE, - $1.50.

POSTAGE to any part of the world PREPAID.

The author of this treatise has made a special study
of Static Electricity, and is an acknowledged master
of the subject. The book embodies the result of
much original investigation and experiment, which
Dr. Atkinson's long experience as a teacher enables
him to describe in clear and interesting language,
devoid of technicalities.

The principles of electricity are presented untram-
meled, as far as possible, by mathematical formulæ,
so as to meet the requirements of a large class who
have not the time or opportunity to master the in-
tricacies of formulæ, which are usually so perplexing
to all but expert mathematicians.

The views expressed in the book are the result of
many years' experience in the class room, the lecture
room and the laboratory, and were adopted only
after the most rigid test of actual and oft-repeated
experiment by the author.

*Copies of the above book will be sent by mail, to any address
in the world, POSTAGE PREPAID, on receipt of price.
Address*

ORIGINAL PAPERS

ON

DYNAMO MACHINERY

AND ALLIED SUBJECTS.

By JOHN HOPKINSON, F.R.S.

Uniform with Thompson's "Lectures on the Electromagnet."

PRICE, INCLUDING POSTAGE, $1.00.

This collection of papers includes all written on electro-technical subjects by the distinguished author, most of which have been epochal in their character and results.

The papers are arranged according to subject. Five papers relate wholly or in part to the continuous current dynamo ; four are on converters and one each on the theory of alternating current machines and on the application of electricity to light-houses.

In the words of the author "The motive of this publication has been that I have understood that one or two of these papers are out of print and not so accessible to American readers as an author who very greatly values the good opinion of American electrical engineers would desire."

PUBLISHED AND FOR SALE BY

THE W. J. JOHNSTON COMPANY, Ltd.

Times Building, New York.

The

Electromagnet,

BY

Prof. SILVANUS P. THOMPSON, D.SC., B.A., M.I.E.E.

A full theoretical and practical account of the properties and peculiarities of electromagnets, together
with complete instructions for designing
magnets to serve any specific purpose.
Published with the express con-
sent and careful revision of
the author.

Cloth. 280 Pages. 75 Illustrations.
Price, $1.00.

LECTURE I.: Introductory; Historical Sketch; Generalities
Concerning Electromagnets; Typical Forms; Polarity; Uses
in General; The Properties of Iron; Methods of Measuring
Permeability; Traction Methods; Curves of Magnetization
and Permeability; The Law of the Electromagnet; Hysteresis;
Fallacies and Facts about Electromagnets. LECTURE II.: Gen-
eral Principles of Design and Construction; Principle of the
Magnetic Circuit. LECTURE III.: Special Designs; Winding of
the Copper; Windings for Constant Pressure and for Constant
Current; Miscellaneous Rules about Winding; Specifications
for Electromagnets; Amateur Rules about Resistance of Elec-
tromagnet and Battery; Forms of Electromagnets; Effect of
Size of Coils; Effect of Position of Coils; Effect of Shape of
Section; Effect of Distance between Poles; Researches of
Prof. Hughes; Position and Form of Armature; Pole-Pieces
on Horseshoe Magnets. Contrast between Electromagnets and
Permanent Magnets; Electromagnets for Maximum Traction;
Electromagnets for Maximum Range of Attraction; Electromagnets of Minimum Weight; A Useful Guiding Principle;
Electromagnets for Use with Alternating Currents; Electro-
magnets for Quickest Action; Connecting Coils for Quickest
Action; Battery Grouping for Quickest Action; Short Cores
vs. Long Cores. LECTURE IV.: Electromagnetism, etc.
    Copies of the above book promptly mailed to any address,
postage prepaid, on receipt of price. Address

RECENT PROGRESS

IN

ELECTRIC RAILWAYS

BEING A SUMMARY OF CURRENT PROGRESS
IN ELECTRIC RAILWAY CONSTRUCTION,
OPERATION, SYSTEMS, MACHINERY,
APPLIANCES, &c., COMPILED

By CARL HERING.

386 pages and 120 illustrations. Cloth,   -   Price, $1.00

This volume contains a classified summary of the
recent literature on this active and promising branch
of electrical progress, with descriptions of new apparatus and devices of general interest.

CONTENTS.

Chapter I.—Historical. Chapter II.—Development
and Statistics. Chapter III.—Construction and Opera-
tion. Chapter IV.—Cost of Construction and Opera-
tion. Chapter V.—Overhead Wire Surface Roads.
Chapter VI.—Conduit and Surface Conductor Roads.
Chapter VII.—Storage Battery Roads. Chapter VIII.
—Underground Tunnel Roads. Chapter IX. –High
Speed Interurban Railroads. Chapter X.—Miscellan-
eous Systems. Chapter XI.—Generators, Motors and
Trucks. Chapter XII.—Accessories.

*Copies of "Recent Progress in Electric Railways," or
of any other Electrical book or books published, will be
promptly mailed to any address in the world,* POSTAGE
PREPAID, *on receipt of the price. Address*

THE W. J. JOHNSTON COMPANY, Ld.,

Times Building, New York.

PRINCIPLES OF

DYNAMO ELECTRIC MACHINES

AND

Practical Directions for Designing and Constructing Dynamos,

By CARL HERING.

Sixth Thousand. 279 pages. 59 Illustrations Price, $2.50.

CONTENTS.

Review of Electrical Units and Fundamental Laws.
Fundamental Principles of Dynamos and Motors.
Magnetism and Electromagnetic Induction.
Generation of Electromotive Force in Dynamos.
Armatures.
Calculation of Armatures.
Field Magnet Frames.
Field Magnet Coils.
Regulation of Machines.
Examining Machines.
Practical Deductions from the Franklin Institute Tests
of Dynamos.
The So-called "Dead Wire" on Gramme Armatures.
Explorations of Magnetic Fields Surrounding Dynamos.
Systems of Cylinder-Armature Windings.
Table of Equivalents of Units of Measurements.

PUBLISHED AND FOR SALE BY

THE W. J. JOHNSTON COMPANY, Ltd.,

TIMES BUILDING, NEW YORK.